Delicious
SEAFOOD RECIPES

Delicious
SEAFOOD RECIPES

Dr. A.D. DHOLAKIA
M.Sc., LL.B. (Sp), Ph. D.
(Retd.) Research Officer and Head
Fisheries Research Station,
Junagadh Agricultural University, Sikka
Gujarat, India

2010
DAYA PUBLISHING HOUSE
Delhi - 110 035

ISBN 9788170359937

Published by : **Daya Publishing House**
A Division of
Astral International Pvt. Ltd.
– ISO 9001:2008 Certified Company –
4760-61/23, Ansari Road, Darya Ganj
New Delhi-110 002
Ph. 011-43549197, 23278134
E-mail: info@astralint.com
Website: www.astralint.com

Laser Typesetting : **Classic Computer Services**
Delhi - 110 035

Printed at : **Chawla Offset Printers**
Delhi - 110 052

PRINTED IN INDIA

Contents

Seer Fish

Bombay Duck

Backti

Ghole

Catfish

Tuna

Eel

Sole

Anchovy

Surmai

Any Fish with White Meat

Introduction

Indian Recipes

In our Indian recipe section you will find a list of recipes that is all about the flavor and the aroma of the meal. These recipes are packed with a wide variety of spices that once combined with each other make a memorable, unique taste like no other. These coconut chutney recipes combine the tastes of grated coconut, mustard seeds, red chilies, and curry leaves for a side dish that can accompany any Indian main dish. Chicken tikka masala that combines the flavors of garlic, ginger, garam masala, cardamom, cumin, paprika, and chili powder for a recipe so flavorful you will be making it again and again. Here you will find traditional Indian food recipes like pannier and pannier with spinach. We also have a recipe for your own garam masala spice recipe so you are sure to get that great traditional Indian flavor in your meals every time. Our garam masala recipes includes the following aromatic spices, bay leaves, cinnamon sticks, cardamom pods, cumin seeds, coriander seeds, peppercorns, cloves, and ground cloves. We also have a recipe for ghee which is a traditional butter type ingredient that is used in traditional Indian cooking. It offers a bold, nutty flavor that regular butter cannot offer. We will also be adding more Indian recipes soon.

In the coastal areas of India, fish and other seafoods are hot favorites. They are either fried, or made as curry or preserved as pickles. They are dried in the sun and preserved using salt for future use also. Generally, fish dishes are healthier than other non-vegetarian foods.

Seafood recipes are so numerous because there are so many flavors that go so well with seafood. There are also many types of seafood such as shrimp, tuna, salmon, catfish, swordfish, cod, and snapper. For a light and refreshing lunch try tomatoes stuffed with tuna. This is an easy recipe using tuna, tomatoes, lemon juice, olive oil, capers, and tomato paste. Poached salmon lying on a hot bed of penne pasta with a thick rich sauce made with dill, tarragon, and lemon flavors. For a spicier take on fish you can use this red curry fishcake recipe. By using your choice of seafood (fish, sardines, shrimp, etc) and covering them in a spicy red curry paste and making a fishcake and frying to perfection. With this zesty baked catfish recipe you will be using tarragon, lemon and pepper seasoning, and a French dressing for a unique taste on catfish fillets.

In this book I have tried to cover different seafoods available. So that one can select recipe from the available seafood in their area. Total 160 delicious seafood recipes including 42 recipe methods are for fresh fish, 43 for prawns/shrimps, 27 from other seafoods like Lobster, Muscle, Clams, Crabs, squids etc and 9 recipes from Dry Fish Products., 5 Pickle and 24 curry preparations are explained. 5 preparations are given for minced seafood meat and 5 preparations are from canned seafood items. I have also given essential information for preparing delicious seafood recipes in appendix. They are:

1. List of 67 various herbs and species used in this book with their Hindi names.
2. Standard weights and measures and their conversion.
3. Glossary of selected cooking items.

Prof. Anshuman D. Dholakia
201 Shashwat Appt. Opp. Sarthi-3 Bungalows
Nr. Vaibhav Laxmi Dham, Memnagar
Ahmedabad – 380 052

Chapter 1

Fresh Fish Products

Seafood can significantly contribute to our nutrient needs, because it is a source of high-quality protein, vitamins and minerals while being low in saturated fat. A well-balanced diet including a variety of fish and shellfish contributes to the health of the heart and to the growth of children. It is also very versatile and seafood recipes are easy to prepare and usually not so much time-consuming. Besides of that seafood is still very healthy.

Everyone knows that fish is a very healthy food. It's low in saturated fat and is a fantastic source of Omega-3 fatty acids, which have been shown to lower the risk of heart disease by lowering cholesterol and decreasing the stickiness of blood platelets. Therefore a diet rich in fish is highly recommended.

Fish is not only healthy but also tasty. There are so many different types of fish you can choose from. Besides of that, you have so many cooking possibilities – you can bake, fry, grill, steam, poach. And it also cooks quickly. But fish can also be eaten uncooked, such as when prepared as sushi and sashimi.

Different dishes from different fresh fish can be prepared as under:

Pomfret

F1) Pomfrets in Arabiata Sauce

Ingredients

1.	Pomfret steaks	8–10
2.	Tomato puree	1 cup
3.	Onion (Chopped)	1
4.	Capsicum (Chopped)	1
5.	Vinegar	1 tsp
6.	Salt	1 tsp
7.	Sugar	1 tsp
8.	Black/white pepper	1 tsp
9.	Oregano	¼ tsp
10.	Thyme	¼ tsp
11.	Basil	¼ tsp
12.	Oil	1 tbsp

Mint leaves for garnishing.

Method

1. Steam pomfret steaks.
2. Prepare the Arabiate sauce by frying onions and capsicum in oil.

3. When they are golden, add tomato puree.
4. Stir and add vinegar, salt, Sugar, Pepper, oregano thyme and basil.
5. Cook on low flame for 5 minutes.
6. Pour over steamed fish just before serving and garnish with mint leaves.

F2) Pomfret Chutney

Ingredients

1.	Medium fish like pomfret cleaned and cut into slice	One
2.	Coconut	100 gm
3.	Coriander leaves	1 cup
4.	Mint leaves	¼ cup
5.	Green chillies	4
6.	Garlic flakes	6
7.	Ginger	1" piece
8.	Chaniajeera powder	1 tsp
9.	Sugar	1 tsp
10.	Juice of one lemon	
11.	Salt and chilli powder	To taste.

Method

1. Grind all the above ingredients except fish to paste.
2. Apply thickly on fish slices.
3. Place each slice separately on Aluminium foil squares.
4. Fold the foil in such a way to form sealed parcels.

5. Put 2 cups of water in a pressure cooker.
6. Place the fish on the grid in the cooker.
7. Fit the lid and bring to pressure rise to maximum and cook for 3 minutes.
8. Reduce the pressure to room temperature.
9. Remove the fish from foil and decorate with lemon slices.

F3) Rache Pomfret

Ingredients

1. Pomfret (Cleaned)	1 large
2. Red Kashmiri chillies	5
3. Cumin	¼ tsp
4. Garlic	1–2 flakes
5. Pepper corns	3
6. Ginger	¼ " piece
7. Tamarind	walnut size ball
8. Vinegar	to taste
9. Salt	to taste

Method

1. Make two slits on both sides of the pomfret and on either side of the back bone.
2. Grind together the chillies, cumin, garlic, pepper corn, tamarind, and ginger in vinegar.
3. Add salt to taste.
4. Stuff the pomfret with masala, on either sides the slit.

5. Secure the fish with string and fry till both the sides are brown.
6. Remove the string and serve.
7. Serves–4.

F4) Stuffed Pomfret

Ingredients

1.	Pomfret	1 large
2.	Coconut	½
3.	Garlic	2 flakes
4.	Ginger	¼" piece
5.	Green chillies	2
6.	Pepper corns	3
7.	Coriander leaves	one large bunch
8.	Tamarind	walnut size ball
9.	Sugar	to taste
10.	Salt	to taste.

Method

Preparation of Chutney

Scrap and grind fine the coconut together with coriander leaves, garlic, ginger, tamarind, green chillies, salt, sugar and pepper corns.

Preparation of Pomfret

1. Leave the pomfret whole and slit on either side.
2. Stuff with the chutney.
3. Secure with string , dot with ghee.

4. Bake in a moderate oven, till done.
5. Remove string and serve garnished with sliced lemon.
6. Serve 4.

F5) Pomfret Rissolves

Ingredients

1.	Pomfret or any fleshy fish	1 large or 2 medium
2.	Tomato (Chopped)	2 (large)
3.	Turmeric	1 tsp
4.	Chilli powder	½ tsp
5.	Ground pepper	½ tsp
6.	Butter	1 tbsp
7.	Salt	½ tsp
8.	Coriander leaves	bunch
9.	Mint leaves	bunch
10.	Onion (chopped)	1 large
11.	Green chillies	2 minced
12.	Lemon juice	1
13.	Potatoes	5 medium and mashed.
14.	Egg	1
15.	Flour	1 tbsp.

Method

1. Boil the fish with the tomatoes and a little water.
2. Mash and set aside.
3. Mix in the turmeric powder, pepper, salt, butter, coriander and mint leaves, onion, green chillies and lemon juice.

4. Add the mashed potato and egg yolk.
5. Beat the egg white and bread crumbs.
6. Fry till done.
7. Serve with the sauce made up of strained stock after boiling the fish, thickened with flour and seasoned with salt and pepper.
8. Serve six.

F6) Pomfret in Mayonnaise Sauce

Ingredients

1.	Pomfret large	1
2.	Tomato chopped	2–3
3.	Ginger chopped	½" piece
4.	Garlic chopped	3 flakes
5.	Butter	2 tbsp
6.	Flour	1 desert spoon
7.	Egg (Yolk only)	1
8.	Milk	1 cup
9.	Mustard	¼ tsp
10.	Pepper powder	½ tsp
11.	Chilli powder	½ tsp
12.	Olive oil	1 desert spoon
13.	Grated cheese	2 tbsp
14.	Salad of tomato, onion and cucumber	

Method

1. Boil the fish with tomatoes, ginger and garlic.
2. Put enough water to just cover the fish.

3. Melt the butter and add the flour gradually with stirring.
4. Beat egg yolk with the seasoning, beat in the milk.
5. Add it to the butter/flour mixture.
6. Cook on a slow fure till thick, stirring all the time.
7. When it cools, add the olive oil and the grated cheese.
8. Serve fish surrounded with the salad.
9. Pour sauce over the whole.
10. Serve hot. It serves 4

Sardin

F7) Spicy Sardines (Talshilele Pedvey)

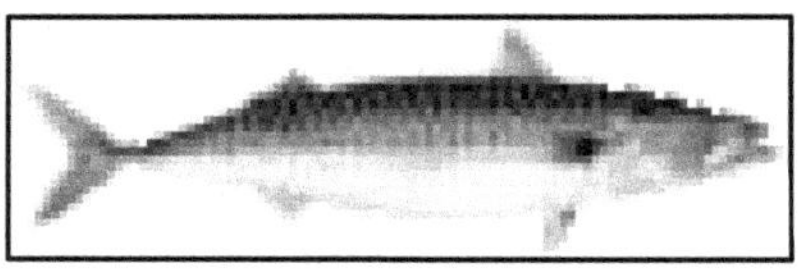

Style: South Indian (Mangalore) Non-vegetarian

Cooking time (approx.): 20 minutes

Ingredients

1. Headless Sardines, insides cleaned	300 gms
2. Coriander powder	1 tsp
3. Turmeric powder	½ tsp
4. Red Chilli powder	1 tsp
5. Asfoetida powder	½ tsp
6. Tarmarind paste	1 tsp
7. Coconut oil	2 tsp
8. Salt	To taste.

Method

1. Wash the sardines well. Rub a litte salt and some turmeric powder on them. Keep aside for about 5 minutes.
2. Heat the oil in a heavy-bottomed pan till hot and put off the flame. Put in the dry spice powders and stir well. Return to a medium-low flame. Put in the tamarind paste, water and salt. Mix well. Add the sardines. Cover and cook on low for about 20 minutes or till the sardines are cooked and coated with the gravy.

Serves

Serve hot with: Appetizing Kokum in Coconut Milk (Sol Kadi) poured over plain boiled rice to 4 Persons

Tips

☆ Coconut oil imparts an exotic aroma to this recipe and is a must traditionally. However, any other cooking oil can be used too.

☆ This curry is supposed to be spicy and tangy. However the amount of chilli powder and tamarind can be reduced or increased as desired.

☆ Canned sardines could also be used instead of fresh ones.

F8) Grilled Sardine

Ingredients

1. Sardines	600 gm.
2. Shallot finely chopped	1
3. Olive oil	splash
4. Garlic crushed	2 cloves

5.	Lemon juice	½ lemon
6.	Lime juice	½ lime
7.	Fresh basil finely chopped	1 spring
8.	Brown sugar	1 tsp
9.	Balsamic vinegar	1 tsp
10.	Poppy seeds	1 tbsp

Method

1. The fish need to be bathed in the marinade for at least an hour, take them from the marinade, grill and eat immediately.
2. Char grilled peppers and couscous is a great accompaniment to this dish.
3. Can't be bothered with the marinade?
4. Simply brush a little olive oil on both sides of the fish then grill; they'll still be simply delicious.
5. Sardines can vary in size so adjust you cooking time accordingly
6. Wash the sardines and soak in cold water for 10 minute.
7. To prepare the marinade simply mix all of the ingredients except half the basil and the poppy seeds together in a shallow dish.
8. Place the sardines in the dish and marinade.
9. Remove the fish from the marinade, place on a tray under the grill and cook for about 12 minutes turning regularly
10. Remove from the grill shower with the remaining basil and poppy seeds then serve

F9) Sauced Sardine

Ingredients

1.	Fresh Sardines	1 kg.
2.	Flour	125 gm
3.	Onion julienne cut	2 big.
4.	Oil	for frying
5.	Olive oil	60 gm
6.	Celery sticks julienne cut	2
7.	Carrot julienne cut	1
8.	White pepper	½ tsp
9.	Cinnamon	little
10.	Parsley	2 springs
11.	Garlic	1 clove
12.	Salt	to taste.

Method

1. Clean the sardines and remove heads.
2. Open the fish flat, wash and dry them and dust with flour.
3. Deep-fry the fish in very hot oil until crisp. Drain on paper towels & season with salt - set aside
4. Heat the extra virgin olive oil in saucepan & fry the onions, carrots & celery until translucent but not browned.
5. Add the vinegar & season with cinnamon and pepper.
6. Boil for 2 minutes, then remove from the heat.
7. Arrange the sardines in layers in a bowl & slowly add the vegetable and vinegar infusion.

8. Marinate for at least 24 hours, then drain & serve garnished with parsley leaves

F10) Crumbed Sardine

Ingredients

1.	Sardines	400 gm
2.	Cake flour	50 gm
3.	Egg	1
4.	Milk	15 ml
5.	Dried bread crumbs	125 gm
6.	Oil	2 tbsp
7.	Salt	coarse little

Method

1. Dust the sardines with flour.
2. Beat the egg and milk together and dip the fish in the mixture.
3. Roll the fish in the breadcrumbs, coating them completely.
4. Deep fry in hot oil, then drain on absorbent paper.
5. Sprinkle the fish with coarse salt and serve them immediately.

Seer Fish

F11) Meen Varuval Receipe Prepared from Seer Fish

In south India, Meen Varuval is prepared from Seer fish. Learn how to make/prepare fried fish by following this easy recipe.

Ingredients

1.	Seer fish steaks	4½ inch
2.	Ground coriander	2½ tbsp
3.	Water	6 tbsp
4.	Vegetable oil for frying	
5.	Fresh lemon juice	5–6 tbsp
6.	Lemon wedges for garnishing	
7.	Salt	2 tsp
8.	Cayenne	2½ tbsp

Method

1. In a skillet, put together cayenne, turmeric, coriander, cumin, salt, lime juice and about 6 tbsp of water.
2. Mix it properly to prepare a thick paste
3. Use this paste to cover both sides of the fish steaks.
4. Marinate the fish steaks for 15 minutes
5. In a frying pan, heat oil.
6. Fry the steaks until they turn brown. Take two steaks at a time.
7. Cook either side for 6-8 minutes, turning once.

8. Place them in a heatproof plate and keep them warm in an oven.
9. Use lemon wedges for garnishing
10. Serve it with rice and naan.

Bombay Duck

F12) Fried Bombay Duck

Ingredients

1.	Fresh Bombay duck	8
2.	Turmeric	½ tsp
3.	Red chilli powder	¼ tsp
4.	Flour	1 cup
5.	Salt	to taste

Method

1. Remove the centre bone of the Bombay duck fish.
2. Flatten and keep under a weight for an hour.
3. Squeeze out all the water.
4. Add the turmeric, salt and chilli powder in this mixture.
5. Pat on both sides.
6. Shallow fry and serve hot.

F13) Battered Bombay Duck

Ingredients

1.	Fresh Bombay duck	8
2.	Turmeric	½ tsp
3.	Red chilli powder	¼ tsp
4.	Egg	2

5.	Bread crumbs	50 gm
6.	Flour	1 cup
7.	Salt	to taste

Method

1. Remove the centre bone of the Bombay duck fish.
2. Flatten and keep under a weight for an hour.
3. Squeeze out all the water.
4. Sprinkle the fish with turmeric, chilli powder and salt.
5. Dip in a beaten egg and bread crumbs.
6. Fry on low flame till done.
7. Serves 4

Backti

F14) Boneless Backti Fish Fry

Ingredients

1.	Bakti fish	500 gm (5 pieces)
2.	Ginger paste	½ tsp
3.	Garlic paste	½ tsp
4.	Lemon juice	1 ½ ysp
5.	Bread crumbs	50 gm
6.	Salt	to taste.

Method

1. Remove scales and intestine of fish.
2. Clean with clean water and make fillet.
3. Marinate all this.
4. Paste it on fish pieces using bread crumbs.

5. Then fry them in a deep frying pan on low flame.
6. Serve hot.

F15) Boneless Backty Orley

Ingredients

1.	Bakti fish	500 gm (5 pieces)
2.	Ginger paste	½ tsp
3.	Garlic paste	½ tsp
4.	Lemon juice	1 ½ tsp
5.	Corn flour	300 gm
6.	Butter	75 gm
7.	Eggs	3
8.	Milk	According to requirement.
9.	Salt	to taste.

Method

1. Mix the ingredients and make paste.
2. Stirr egg and butter.
3. Mix with the paste.
4. Now dip the pieces of fishes in it.
5. Fry it in deep frying pan.

Ghole

F16) Ghole (Pabda) Fish Jhal (Bengali recipes)

Ingredients

1.	Ghole (Pabda) fish	500 gm.
2.	Mustard	100 gm
3.	Tomato pieces	2
4.	Green chillies	2

5.	Turmeric powder	25 gm.
6.	Chilli powder	1 tsp
7.	Coriander leaves	1 ½ tsp
8.	Mustard oil	100 gm
9.	Garlic	3 pods
10.	Ginger	1' piece
11.	Dry red chilli whole	2
12.	Paste of mustard, garlic.	

Method

1. Take a small Karchi and heat it for a second.
2. Then put mustard oil and slightly heat it.
3. Put dressed fish pieces one by one and fry slowly.
4. Fry the fish till it becomes light brown.
5. Put all the paste and stir it for a few second.
6. Then put ½ a cup of water to boil.
7. Then put salt and ½ tsp turmeric powder when water dries up.
8. Garnish with coriander leaves and serve hot.

F17) Basil Broiled Grouper

Ingredients

1.	Skinless Grouper fillets	900 gm
2.	Butter	4 tbsp
3.	Soy sauce	1 tbsp
4.	Pepper	1/8 tsp
5.	Lemon juice	1 tbsp
6.	Basil butter	½ cup

Method

1. Prepare BASIL BUTTER: Combine ½ cup margarine or butter with 1 tablespoon dried basil and mix well.
2. Sprinkle fish with salt and pepper.
3. Place fish in a single layer on a well-greased broiler pan, 12 x 8 x 2 inches.
4. Combine margarine, soy sauce and lemon juice.
5. Brush half the sauce on top side of fish.
6. Broil about 4 inches from source of heat for 4 to 5 minutes.
7. Turn carefully, brush with remaining sauce.
8. Broil 4 to 5 minutes longer or until fish flakes easily when tested with a fork.
9. Top servings with equal amounts of Basil Butter.
10. Makes 6 servings.

Catfish

F18) Catfish Fingers with Spicy Dipping Sauce

Ingredients

1. Catfish fillets	2
2. White corn flour finely ground	2 cups
3. Black pepper freshly ground	½ tsp
4. Peanut oil	1 ½ quarts
5. Salt	2 tsp

Sauce Ingredients

1. Vinegar	¾ cup
2. Black pepper coarsely ground	½ tsp

3.	Shallot finely chopped	1
4.	Garlic finely chopped	2 cloves
5.	Scallions	2
6.	Parsley	1 tbsp

Method

1. Catfish fillets cut into 1 inch wide strips.
2. Catfish strips carefully with salt and pepper. Set aside.
3. Combine cornmeal, salt and pepper.
4. Dredge catfish strips in cornmeal mixture and fry in 2 inches of peanut oil heated to 375° F.
5. Fry until golden brown, about 5 minutes.
6. Drain catfish strips on crumpled paper towel and sprinkle lightly with salt.

F19) Fettuccini with Catfish

Ingredients

1.	Catfish fillets	3
2.	Butter	2 tsp
3.	Onion	3
4.	Black pepper coarse ground	2 tsp
5.	White wine	½ cup
6.	Milk	¾ cup
7.	Whipping cream	¾ cup
8.	Nutmeg	¼ tsp
9.	Lemon juice	1 tsp
10.	Fettuccine	8 ounces
11.	Salt	to taste

12.	Parsley chopped	1 tsp
13.	Chives	1 tsp

Method

1. Cut catfish fillets in thin strips.
2. In a heavy skillet, over medium heat, melt butter.
3. Cook the onions and pepper for 1 minute.
4. Add the wine and catfish strips, stirring gently to coat catfish on all sides.
5. Add the milk, cream, nutmeg and salt, continuing to stir.
6. Add the lemon juice and stir constantly while the sauce thickens.
7. Check the seasonings and adjust if necessary.
8. Serve immediately tossed with the fettuccine.
9. Sprinkle each serving with parsley or chives

F20) Spicy Grilled Catfish

Ingredients

1.	Catfish fillets	450 gm
2.	Olive oil	2 tbsp
3.	Lime juice	¼ cup
4.	White wine	1 cup
5.	Dry mustard	2 tbsp
6.	Chilli powder	2 tbsp
7.	Pepper	2 tsp
8.	Fresh cilantro chopped	1.2 cup
9.	Salt	to taste.

Method

1. Prepare a grill or preheat the broiler.
2. Mix olive oil, lime juice, wine, mustard, chili powder, pepper, cilantro and salt in a medium bowl.
3. Transfer half of the marinade mixture to another bowl and reserve for basting catfish.
4. Add catfish to the first bowl; marinate for 15 minutes.
5. Drain fillets and discard marinade.
6. Place fillets on an oiled grill rack or broiler pan rack.
7. Grill or broil 4 inches from the heat source, basting with reserved marinade, for about 3 minutes on each side, or until fish flakes easily when tested with a fork.

F21) Baked Catfish

Ingredients

1. Catfish fillets	250 gm
2. Butter	125 gm
3. Dry white wine	125 ml
4. Lemon juice	1 tbsp
5. Garlic finely chopped	1 tsp
6. Cilantro chopped	1 tsp
7. Pepper	to taste
8. Salt	to taste
9. Paprika	1 tsp

Method

1. Season fish with salt and pepper.
2. Place fish in oval casserole dish.
3. Melt 4 ounces butter in small sauce pan.
4. Add wine and remaining ingredients, except the paprika.
5. Simmer about 2 minutes. Remove from heat and spoon over fish.
6. Now sprinkle paprika over fish.
7. Bake fish in 375 degree oven for 10 to 12 minutes, or until the fish is done.
8. Serves 2

Tuna

F22) Tuna Rice Casserole

Ingredients

1. Canned light Tuna	200 gm
2. Mushroom soup	325 gm

3.	White rice	1 cup
4.	Butter	1 tsp
5.	Bread crumbs dry	1 ½ tbsp
6.	Cheddar cheese	60–70 gm

Method

1. Cook rice following package directions.
2. Stir soup into rice.
3. Mash tuna and stir into rice mixture.
4. Spread butter over inside of 1 ¾-quart casserole; sprinkle bread crumbs over butter.
5. Pour tuna-rice mixture into casserole and sprinkle with grated cheese.
6. Bake on middle shelf of preheated 375-degree oven 30 minutes.
7. Serves 4

F23) Tuna Sandwiches

Ingredients

1.	Canned tuna drained and flaked	200 gm
2.	Swiss cheese	½ cup
3.	Celery chopped	½ cup
4.	Onion finely chopped	2 tbsp
5.	Mayonnaise	¼ cup
6.	Sour cream	¼ cup
7.	Pepper	to taste.
8.	Bread slices	16

Method

1. In a bowl combine the tuna, cheese, celery, onion, mayonnaise, sour cream, and pepper to taste.
2. Spread on the bread and serve at once.
3. 8 sandwiches will be ready.

F24) Grilled Tuna Sticks

Ingredients

1.	Tuna steaks	250 gm
2.	Fresh lemon juice	¼ cup
3.	Virgin olive oil	1 tbsp
4.	Fresh onion chopped	2 tsp
5.	Lemon zest	1 tbsp
6.	Freshly ground pepper	1 tsp
7.	Salt	¼ tsp

Method

1. Tuna steaks are marinated in a garlic-lemon-oregano marinade, then grilled.
2. Prepare a charcoal grill.
3. Meanwhile, rinse the tuna, pat dry, and place in a shallow glass or ceramic dish.
4. In a small bowl, whisk the remaining ingredients.
5. Pour over the fish and cover with plastic wrap.
6. Marinate for 30-45 minutes in the refrigerator.
7. When the fire is ready, drain the tuna and place it in a fish basket. Grill the steaks 4" from the heat, 5 minutes per side or until tuna is done.
8. Baste with remaining marinade if desired.

F25) Tuna Patties

Ingredients

1.	Tuna canned drained	900 gm
2.	Egg beaten	2
3.	Onion finely chopped	¼ cup
4.	Ritz crackers crushed to crumbs	¾ sleeve

Method

1. Mix all ingredients together well.
2. Preheat a skillet with some oil over medium high heat.
3. Form mixture into 6 to 8 patties and fry until golden and crispy on the outside and heated through.

Eel

F26) Backed Eel

Ingredients

1.	Eel (large variety)	900 gm
2.	Olive oil	½ cup
3.	Wine vinegar	2 tbsp
4.	Bay leaves	4
5.	Pepper	½ tsp
6.	Bread crumbs	1 tbsp
7.	Salt	1 tsp

Method

1. Mix together oil, vinegar, bay leaves, salt, pepper and bread crumbs.
2. Marinate eel pieces in this mixture 3 hours, turning frequently.
3. Place eel pieces on skewers, alternating with bay leaves.
4. Place skewers in greased baking dish, and bake in moderate oven (375), 30 minutes, turning often.
5. Brush with remaining marinade while cooking.
6. Serves 4

Sole

F27) Backed Lemon Sole

Ingredients

1.	Sole fillets	450 gm
2.	Vegetable oil	¼ cup
3.	Fresh lemon juice	¼ cup

4.	Bread crumbs	3 tbsp
5.	Grated cheese	3 tbsp
6.	Paprika	½ tsp

Method

1. In a small bowl combine vegetable oil and lemon juice with whisk; pour into 13 x 9 x 2 inch glass baking dish.
2. Rinse Fillets in cold water, pat dry, add to baking dish and turn Fillets over to coat with oil-lemon juice mixture.
3. Sprinkle Fillets evenly with dry bread crumbs, then with grated Parmesan cheese.
4. Dust with paprika. Bake at 350 F 15 minutes or until fish flakes easily with a fork.
5. May also be done in microwave on high for about 3-4 minutes.
6. Serves 4

F28) Pescado Brasado (Mexican Dish)

Ingredients

1.	Sole fish	2 pieces
2.	Bread slices	2
3.	Green rice	100 gm
4.	Potato turned	100 gm
5.	Red chillies	2
6.	Thyme	1 tsp
7.	Tomato purees	50 gm
8.	Red Chilli paste	10 gm

9.	Pepper	1 tsp
10.	Oregano	½ tsp
11.	Salt	to taste

Method

1. Grill fish on a hot plate.
2. Fry bread slices till they are golden brown.
3. Make green rice with the help of green leaf.
4. Make turned potato, blanch it.
5. Saute it with salt and oregano.
6. Heat fat in a pan, put tomato puree, chilli paste, red chilli whole, thyme, salt pepper.
7. Cook fish in this sauce for a few minutes.
8. Set rice in one side of plate and potato on other.
9. Set fish on bread slice with sauce.
10. Garnish it with olive, parsly and lemon star.

F29) Backed Fillets of Sole

Ingredients

1.	Sole fish (fresh or frozen)	900 gm
2.	Fresh lemon juice	¼ cup
3.	Canned mushroom	60 gm
4.	Butter	2 tbsp
5.	Bay leaf	1 small
6.	Pepper	1/8 tsp
7.	Grated Onion	1 tbsp
8.	Flour	2 tbsp
9.	Egg yolks	2

10.	Frozen spinach	100 gm
11.	Water	¼ cup

Method

1. Preheat oven to 350 F. Combine lemon juice, water, bay leaf, salt and pepper, and marinate fish in mixture for 1 hour.
2. Then drain liquid into a bowl, add mushroom liquid and enough water to make 1 ½ cups.
3. Melt butter, add grated onion and flour and stir in liquid slowly.
4. Cook until smooth and thick.
5. Add spinach for 2 minutes, drain and season, arrange in a greased baking dish.
6. Top with fish and sauce.
7. Dot with butter. Bake at 350 F for 30 minutes.
8. Serves 4

Anchovy

F30) Anchovy Butter

Ingredients

1.	Anchovy fillets	8
2.	Lemon juice	1
3.	Fresh parsley	1 tbsp
4.	Butter	1 cup

Method

1. Mash anchovy fillets with the lemon juice in a mortar until all the juice has been incorporated.

2. Mix in the chopped fresh parsley.
3. Add the mixture to the butter and beat well to form a smooth paste.
4. Pack into a small crock and refrigerate for at least 1 hour before using.

F31) Anchovy Cheese Dip

Ingredients

1.	Anchovy paste	1 tbsp
2.	Cream cheese	250 gm
3.	Chives, chopped	1 tbsp
4.	Green olive minced	2 tbsp
5.	Lemon juice	1 tsp
6.	Milk	1 tbsp

Method

1. Combine all ingredients.
2. Beat at medium speed of electric mixer until light and fluffy.
3. Chill until ready to serve.
4. Serve with corn chips or crackers.

F32) Anchovy, Fried Genovese Style

Ingredients

1.	Fresh Anchovy	550 gm
2.	Eggs	3
3.	Olive oil	3 cup
4.	Garlic	3 cloves
5.	Parsley finely chopped	¼ cup

6.	Bread crumbs finely chopped	4 cup
7.	Salt	to taste
8.	Black pepper	to taste
9.	Lemon	2

Method

1. Heat oil to 375 degrees and maintain heat.
2. Scale and gut anchovies, leaving heads and tails intact.
3. In a large mixing bowl, mix eggs, garlic and parsley until well blended.
4. Dip fish 3 or 4 at a time into egg mixture and then into bread crumbs.
5. Shake well to remove excess egg-crumb coating and fry fish 3 to 4 at a time until crisp and golden brown, about 1 to 2 minutes.
6. Remove and drain on a paper towel.
7. Serve immediately with salt, pepper and lemon wedges.
8. This recipe yields 4 servings as antipasto.

Surmai

F33) Surmai in Lemon Juice

Ingredients

1.	Thick slice of Surmai.	500 gm
2.	Lemon juice	1 tbsp
3.	Red chilli powder	1 tbsp
4.	Black pepper powder	½ tsp
5.	Turmeric powder	½ tsp

6.	Cumin seeds	½ tsp
7.	Coriander seeds	½ tsp
8.	Onion chopped	1 small
9.	Garlic chopped	2 tsp
10.	Ginger chopped	2 tsp
11.	Curry leaves	4
12.	Salt	to taste
13.	Cucumber and tomato	to garnish

Method

1. Mix ingredients no 3 to 12 and grind to a fine paste add water as required.
2. Apply a little salt and lemon juice to the washed fish slices.
3. Keep for 5-10 minutes.
4. Wash and drain well.
5. Marinate the fish slices in the ground paste and keep preferably refrigerated for about an hour.
6. Heat a griddle on medium flame and smear a little oil on it.
7. Place the slices in batches of 2 or 3 depending upon the size of the griddle.
8. Do not overcrowd the griddle.
9. Pour a little oil around each fish slice with a teaspoon and let it roast on a medium flame till aromatic and crisp.
10. Flip and repeat the procedure for the other side.
11. Serve hot.

Tips

- ☆ Any white firm flesh fish can be used instead of king fish (surmai).
- ☆ To make this recipe a low fat one, the marinated fish slices can be grilled with minimum or no oil.

Garnishing

1. On a bed of onion rings, cucumber and tomato slices garnished with lemon wedges and accompanied with a tall glass of chilled wine.

Mackeral

F34) Mackeral Fish Kofta (Bengali Recipe)

Ingredients

1.	Mackerel fillets, skinned	500 gm
2.	Fresh coriander chopped	2 tbsp
3.	Poppy seeds ground	2 tsp
4.	Onion, finely chopped	1 small

5.	Garlic crushed	2 large cloves
6.	Egg, beaten	1
7.	Ginger, finely chopped	½ inch
8.	Garam masala	2 tsp
9.	Oil	for deep–frying
10.	Red chilli	¼ tsp
11.	Salt	to taste.

Sauce

- ☆ 3 large cloves garlic, chopped
- ☆ 2.5 cm/1 in ginger root, chopped
- ☆ 1 onion, chopped
- ☆ 75 ml/5 tbsp oil
- ☆ large pinch of asafetida
- ☆ 5 ml/1 tsp mustard seeds
- ☆ 2 bay leaves
- ☆ 1 small onion, finely chopped
- ☆ 5 ml/1 tsp ground coriander (cilantro)
- ☆ 5 ml/1 tsp ground roasted cumin
- ☆ 2.5 ml/½ tsp garam masala
- ☆ 2.5 ml/½ tsp ground red chilli
- ☆ 2.5 ml/½ tsp ground turmeric
- ☆ salt
- ☆ 225 g/8 oz canned tomatoes
- ☆ 150 ml/¼ pt/2/3 cup natural (plain) yoghurt
- ☆ 250 ml/8 fl oz/1 cup water

Garnish

☆ 15 ml/1 tbsp chopped fresh coriander (cilantro)

Method

1. Bake the fish in a preheated oven at 240°C/475°F/ gas mark 9 for 20 minutes.
2. Mix together the coriander, poppy seeds, onion, garlic, ginger, egg, garam masala, chilli and salt and add to the baked fish.
3. Divide the mixture into 18 equal portions and roll them into rounds. Heat the oil and fry the koftas until golden brown.
4. Make the sauce while the fish is cooking.
5. Blend the garlic, ginger and onion to a smooth paste in a blender or food processor.
6. Heat the oil and fry the asafetida and mustard seeds until the seeds start crackling.
7. Add the bay leaves and small onion and fry until golden brown.
8. Add the blended paste and fry until golden brown.
9. Stir in the ground spices, salt and tomatoes and cook until all the liquid has been absorbed and the oil appears on the surface.
10. Stir in the yoghurt and cook until all the liquid has evaporated.
11. Pour in the water, bring to the boil then simmer for 5 minutes.
12. Pour the hot sauce over the koftas, sprinkle with the garnish ingredients.
13. Serve hot with rice, chapattis, raita and a potato dish.

Any Fish with White Meat

F35) Kerala Fish Fry (Karimeen Pollichathu)

This is a traditional Kerala (southern part of India) favorite. Fish pieces are sauted with onions and spices to give a spicy semi-dry preparation.

Style: Kerala Non-Vegetarian

Cooking time (approx.): 14 minutes

Ingredients

1.	Skinless firm white fish slices	500 gm
2.	Finely sliced garlic	1 tsp
3.	Ginger juliennes	1 tsp
4.	Green chillies	4 Nos.
5.	Onion (Finely chopped)	1 (medium)
6.	Red chilli powder	1 tsp
7.	Turmeric powder	½ tsp
8.	Tomato chopped	1 small
9.	Curry leaves	1 spring
10.	Oil	2 tbs
11.	Water	2 cups
12.	Salt	to taste

Method

1. Heat the oil in a heavy-bottomed pan till hot. Add the garlic, ginger, green chillies and onions. Saute for about 4 minutes.
2. Add the turmeric powder, red chilli powder, curry leaves and salt. Saute briefly.
3. Add the water and raw mango slices/chopped tomato.

4. Bring to a boil. Drop in the fish and mix gently.
5. Cover and cook on low for about 10 minutes or till the fish are cooked and the gravy is semi-dry.

Tip

☆ For a sour taste, either vinegar, lime juice or tamarind paste can be used in place of raw mango or tomato.

Serve hot with: warm Lacy Rice and Coconut Pancakes (Appam) and/or steamed rice to 4 persons.

F36) Fish Roll

Ingredients

1. Fish	500 gm (7 pieces)
2. Prawns	50 gm
3. Ginger paste	2 tbsp
4. Garlic paste	1 tsp
5. Lemon juice	3 tsp
6. Egg (Boiled and sliced)	2
7. Onion Chopped	200 gm
8. Maida	300 gm
9. Coriander leaves	bunch
10. Salt	to taste.

Method

1. Boil fish and take out only flesh.
2. Similarly boil prawns and remove shell.
3. Heat oil in a bowl.
4. Add all the ingredients to it and stir well.

5. Now with the help of paste, roll maida base roti.
6. Mix masala and boiled fish and prawns.
7. Stuff in a roti and fry.
8. Decorate with tomato and sliced eggs.
9. Serve hot.

F37) Boneless Fish Orley

Ingredients

1. Fish	500 gm (7 pieces)
2. Prawns	50 gm
3. Ginger paste	2 tbsp
4. Garlic paste	1 tsp
5. Lemon juice	3 tsp
6. Coriander leaves	according to taste
7. Onion	200 gm

For preparing paste

1. Corn flour	300 gm
2. Butter	75 gm
3. Eggs	3 Nos.

Milk according to requirement

Method

Mix the ingredients and make a paste. Now dip the pieces of fishes in it. Fry it in a deep frying-pan.

F38) Fish Stew

Ingredients

1. Dressed fish pieces	500 gm
2. Oil	2 desert spoon

3.	Sliced onion	¼ cup
4.	Ginger cut lengthwise	3 Nos
5.	Garlic	8 pods
6.	Curry leaves	few
7.	Butter	1 tsp
8.	Maida	1 desert spoon
9.	Milk	½ cup
10.	White pepper coarsely ground	½ tsp
11.	Vinegar	1 desert spoon
12.	Powdered salt	½ tsp

Method

1. Fry the maida in butter. Add milk and stir continuously.
2. Filter this and keep it on flame till attains a syrupy consistency.
3. Mix coarsely powdered pepper to the syrup to make white sauce ready.
4. Fry the second ingredients no 3 to 6 in oil and keep aside.
5. Cook the fish in the remaining oil with enough water, vinegar and salt
6. Add the second ingredients to the cooked fish.
7. Add white sauce, and when the gravy thickens remove it from fire.
8. The stew is seasoned with cooked carrot, beans and serve hot.

F39) Fish Thorn

Ingredients

1.	Fish	500 gm
2.	Coconut (Scrapped)	½
3.	Green chillies	5–10 Nos
4.	Onion (Sliced)	50 gm
5.	Ginger	1" piece
6.	Pepper powder	½ tsp
7.	Mustard	1 tsp
8.	Coconut oil	100 gm
9.	Black gram Dal	1 tsp
10.	Salt	to taste.

Method

1. Cook fish and remove skin and bones and mash.
2. Fry mustard seed and dal till the mustard burst.
3. Then add pepper powder, sliced onion, ginger and green chillies.
4. Fry till it is brown.
5. Add mashed fish and scrapped coconut and salt to taste.
6. Fry for few minutes till water evaporates.

F40) Fish in Coconut Gravy

Ingredients

1.	Fish pieces	450 gm
2.	Fenugreek powder	¼ tsp
3.	Red chilli	7
4.	Turmeric	1" piece

5.	Cumin seeds	¼ tsp
6.	Scrapped coconut	¾ cup
7.	Garlic cloves	5
8.	Small onion	6–7
9.	Split green chillies	7
10.	Chopped ginger	½ tsp
11.	Curry leaves	few
12.	Coconut oil	¼ cup
13.	Tamarind water and lime juice	as per taste
14.	Salt	to taste.

Method

1. Grind small onion to a paste and keep aside.
2. Grind ingredients 3, 4, 5, and 7 to form paste.
3. Grind the coconut also separately.
4. Fry the three ground pastes one after another in hot oil until golden brown.
5. Add lime juice, tamarind water and salt according to taste.
6. Put the fish pieces into this and allow boiling.
7. Add ingredients 9–11 and cook on low flame with enough water for 10 minutes.
8. Sprinkle ground fenugreek powder on the curry

F41) Fish Pappas

Ingredients

1.	Small pieces of fish	500 gm
2.	Oil	2 desert spoon (dsp)

3.	(a) Mustard	½ tsp
	(b) Fenugreek	2 pinches
4.	(a) Onion chopped lengthwise	¼ cup
	(b) Ginger cut length-wise	1 tsp
	(c) Small cloves of garlic	12
5.	(a) Coriander powder	1 dsp
	(b) Chilli powder	1 tsp
	(c) Turmeric powder	½ tsp
6.	(a) water	1 cup
	(b) Malabar tamarind	4 pieces
7.	(a) Curry leaves	few
	(b) Powdered salt	1 tsp
8.	(a) Cow's milk	½ cup
	(b) Maida	1 tsp

Method

1. Mix the (5) ingredients with a little water and make into a paste.
2. Fry mustard and fenugreek in oil, to it add ingredients 4 and fry till golden brown.
3. Saute the masala paste in the above oil.
4. Add the (6) ingredients and allow to boil.
6. To it add fish pieces and curry leaves.
7. When it boils add salt.
8. When the gravy thickens, simmer the flame and add the mixture of maida, milk and stir.
9. Boil for 5 minutes.
10. Serve hot.

F42) Fish Fry

Ingredients

1. Fish	200 gm
2. Onion	50 gm
3. Turmeric powder	½ tsp
4. Red chillies	10 Nos.
5. Ginger	10 Nos.
6. Coconut oil	100 gm
7. Salt	to taste

Method

1. Grind nicely the turmeric chillies, onion, ginger and salt.
2. Paste the masala on the cleaned and sliced fish.
3. Keep it for 20–30 minutes.
4. Fry it in oil till it is brown.

Chapter 2

Prawns Recipes

Prawn is very popular seafood and is extensively eaten in India. Prawn is a major dish in Bengal and other coastal

areas of India. Indian prawn recipes are very popular. Try our range of easy prawn recipes.

No wonder, shrimp recipes are simply delicious !!! And the cuisines of different cultures have recipes using shrimps.

The meat of shrimps has a mild, sweet flavor and a satisfying texture, and the different sizes can be substituted for each other. In the list below you can find typical recipes from different countries and regions (for instance, Chinese, Italian, Spanish, Brazilian, etc.) that use shrimps as main ingredient. Therefore you can expect recipes that combine shrimps with a wide range of ingredients like curry, coconut, garlic, tomatoes, etc. Our shrimp recipes list comes in different variations, such pasta, risotto, paella, pizza, salad, and many others.

Prawns are one of those expensive and yet important stuffs for Chinese New Year. However, only the Cantonese go for prawns because prawn is call 'har' in Cantonese and it rhymes with the sound of laughter of hahaha. Meanwhile, do you know that Hokkien in tradition, do not like prawns because of we say 'siang kah heh anneh' or 'like prawns' when refer to something dead or useless.

If your prawns are fresh, a steam with some ginger, Chinese cooking wine and salt will be just fine. However, many homes probably kept the prawns for months already. So this may not be a good idea.

Important thing to note before cooking prawns is to smell the fresh ones. If they smell bad, keep those for curries. It is smells of the sea, just plain frying with a bit of salt is good as they are.

At home, it is best done in simple recipes because prawn is tasty on its own without any special flavourings.

P1) Scampial Tarrazzo

This delicious dish is reminiscent of hot, exotic settings perhaps a terrace overlooking the Adriatic. Complete the scenario with a crisp, dry white wine–a soave would be the ideal choice.

Ingredients

1.	Fresh peeled scampi	1 kg
2.	Butter	2 tbsp
3.	Olive oil	4 tbsp
4.	Garlic	2 cloves (finely chopped)
5.	Lemon juice	3–4 tbsp
6.	Parsley	6 tbsp (freshly chopped)
7.	Capers	2 tbsp (drained)
8.	Boiled rice to serve	
9.	Garnishing	
10.	Lemon slice	

Method

1. Pat the scampi dry with absorbent paper.
2. Heat the oil in a large, heavy based frying pan and cook the scampi very gently over low flame, turning them now and then, for 6 minutes until it is almost cooked.
3. Add the butter; stir the garlic, lemon juice, parsley and capers into the frying-pan and cook, stirring for further 1–2 minutes.
4. Spoon the scampi and cooking juice over a bed of plainly boiled rice arranged on a serving platter.
5. Garnish the edge of platter with lemon slice and serve immediately.

P2) Prawn Cocktailed Salad

Ingredients

1.	Prawns	250 gm
2.	Cabbage	1 cup
3.	Thousand Island Dressing	½ cup

Method

1. Boil prawns.
2. Keep aside.
3. Layer cabbage (chopped) on a dish.
4. Arrange prawns on it. Pour Thousand Island dressing on top, chill and serve.

P3) Prawn Baffad (Goan Dish)

Ingredients

1.	Prawns (Shelled)	250 gm
2.	Kashmiri chilies	8
3.	Cumin	½ tsp
4.	Peppercorns	12
5.	Flakes garlic	2–3
6.	Ginger	½″ piece
7.	Turmeric	1 tsp
8.	Onion (medium)	2 (Chopped)
9.	Tomato large	1 (Chopped)
10.	Salt	to taste
11.	Tamarind and vinegar	to taste

Method

1. Grind together the chillies, cumin, peppercorns garlic and ginger and mix in the turmeric powder.
2. Fry the onions and add the tomato and prawns.
3. Fry well. Next, add the ground massals and the tamarind and vinegar.
4. Cover and cook till the prawns are done.
5. Serves 4 persons.

P4) Chilli Prawns

Chilli Prawn by following method is easy recipe.

Ingredients

1.	Prawns	2 cups
2.	Onions	2

3.	Capsicums	2
4.	Pepper	2tsp
5.	Soy sauce	1tbsp
6.	Spring onion	2tbsp
7.	Maida	4tbsp
8.	Corn flour	4 tbsp
9.	Garlic flakes	2
10.	Eggs	2
11.	Ajunomoto	1 pinch
12.	Salt	to taste
13.	Pepper and sugar	to taste
14.	Oil	for frying

Method

1. Chop the onions, capsicum, green chillies, garlic and ginger.
2. Clean the prawns. Mix maida, corn flour, eggs, salt, pepper and sugar to make a batter. Heat the oil.
3. Dip the prawn pieces in batter. Fry them till reddish brown in colour.

4. Drain and keep it aside. Heat the oil.
5. Add chopped ginger, garlic, chopped vegetables and green chillies one by one. Toss them well.
6. Now add prawns.
7. Mix salt, sugar, pepper, ajinomoto and soy sauce.
8. Toss the mixture by adding water. Once the mixture is half cooked, add spring onions. Mix it well.
9. Chilli Prawn is ready.

P5) Butterfly Prawn Recipe

Butterfly Prawns is a famous Chinese recipe. Learn how to make/prepare Butterfly Prawns by following this easy recipe.

Ingredients

1.	Large prawns	12
2.	Chinese wine or dry sherry	1tsp
3.	Garlic & clove crushed	1
4.	Light soy sauce	2tbsp
5.	Egg beaten	1
6.	Fresh ginger finely grated	½ tsp

7. Corn flour	½ cup
8. Bread crumbs	for coating
9. Peanut oil	for deep frying
10. Salt	to taste.

Method

1. Shell the prawns and remove the tails.
2. From the interior curve, cut the meat leaving the thin layer on the back.
3. Mix soy sauce, garlic, wine, salt & ginger.
4. Dip prawns into corn flour.
5. Now dip the prawns into the beaten egg.
6. Use breadcrumbs for coating.
7. Fry the prawns properly until they become crispy.

P6) Coconut Prawn Gravy Recipe

Coconut Prawn Gravy is a tasty Malabar recipe. Learn how to make/prepare Coconut Prawn Curry by following this easy recipe.

Ingredients

1. Prawns	500 gm
2. Onions	2
3. Tomato	2
4. Fennel seeds	1 pinch
5. Turmeric powder	1 tsp
6. Chilli powder	2 tsp
7. Coriander powder	3 tsp
8. Green Chillies (chopped)	4

9.	Garlic flakes (crushed)	3
10.	Coconut oil	2 tbsp
11.	Tamarind paste	2 tbsp
12.	Water	2 cups
13.	Curry leaves	
14.	Salt	to taste

Method

1. Chop the onions and tomatoes.
2. Wash and clean the prawns. Heat 1 tsp coconut oil.
3. Heat 1 tsp coconut oil.
4. Fry grated coconut, fennel seeds and chopped onions till brown. Blend it well with water.
5. Roast the turmeric powder, chilli powder and coriander powder. Mix the roasted powders with chopped tomatoes, garlic, blended coconut, salt and tamarind paste.
6. Now add 2 cups water and cook till the mixture turns thick. Add prawns and cook till done.
7. Mix green chillies. Heat the oil. Fry curry leaves.
8. Adorn with fried curry leaves. Coconut prawn gravy is ready.

P7) Garlic Prawns Recipe

Garlic Prawns is a classic recipe, garlic gives a new flavor to it. Learn how to make/prepare Garlic Prawn by following this easy recipe.

Ingredients

1.	Prawns large shelled	500 gm

2.	Garlic crushed	4 tbsp
3.	Butter	5 tbsp
4.	Marjoram	1 tsp
5.	Onion big, sliced	1
6.	Chilli and lime dressing	5 tbsp

Freshly cracked black pepper and Ground Sea salt to taste.

Method

1. Heat pan on stove and melt butter in it.
2. Combine onion and garlic with it, cook for about 30 seconds on high flame.
3. Add prawns to it, mix well.
4. Combine Chilli and lime salad dressing or lemons or lime juice with it.
5. Sprinkle marjoram over the prawns.
6. Add black pepper and sea salt to it.

P8) Golden Fried Prawn Recipe

Golden Fried Prawns are best complemented with mint-coriander chutney. Learn how to make/prepare Golden Fried Prawns by following this easy recipe.

Ingredients

1.	Prawns (King size)	8
2.	Ajinomoto	¼ tsp
3.	Baking powder	½ tsp
4.	Corn flour	100 gm
5.	White pepper	¼ tsp
6.	Oil	2 tsp
7.	White flour (Maida)	50 gm
8.	Egg whites	2
9.	Red wine or sherry	2 tsp
10.	Oil	for frying
11.	Salt and pepper	to taste
12.	Water	½ cup

Method

1. Wash and chop the prawns into pieces.
2. Mix salt, wine, black pepper and prawns properly and leave this mixture for 20 minutes.
3. Take a bowl and prepare a mixture of corn flour, white flour, egg whites, ajinomoto, baking powder, white pepper, water and oil.
4. Stir it until the mixture thickens.
5. Now dip the prawns in this mixture.
6. Fry prawns properly. Golden fried prawn is ready.

P9) Prawn Balchao Recipe

Prawn Balchao is a traditional spicy dish of Goa. Learn how to make/prepare Prawn Balchao by following this easy recipe.

Ingredients

1. Prawns	500 gm. (Headless, deveined, shelled, and cleaned)
2. Tomato	4 (finely chopped)
3. Green chillies	2 (Chopped)
4. Curry leaves	1 spring
5. Onions	4 (Finely chopped)
6. Sugar, salt and vinegar	to taste
7. Oil	4 tbsp

For making paste

1. Ginger–garlic paste	
2. Dried Red chillies	
3. Vinegar	½ tsp
4. Turmeric powder	½ tsp
5. Cumin seeds	2 tsp
6. Mustard seeds	1 tbsp
7. Black peppercorns	1 tsp

Method

1. Mix a pinch of turmeric powder and some salt.
2. Rub the cleaned prawns with this mixture. Leave aside for five minutes.

3. Wash the prawns well and drain.
4. Apply the above prepared fine paste into the prawns and leave aside. Heat oil in a large frying pan and add in the curry leaves.
5. Add finely chopped onions and fry on medium flame till the onions turn light brown.
6. Add chopped tomatoes and saute till the oil starts leaving the sides of pan.
7. Now add the prawns and marinade and fry briefly.
8. Stir occasionally. Add green chillies, sugar, salt and vinegar to taste.
9. Mix well. Put a lid and simmer on low flame for about ten minutes. Remove from the heat when prawns are well cooked.
10. Serve with bread slices or chilled Kokum in Yoghurt.

P10) Prawn Cutlets Recipe

Prawn Cutlets is a classic recipe, prepared by dipping prawns into the egg mixture. Learn how to make/prepare Prawn Cutlet by following this easy recipe.

Ingredients

1. Prawns large size	12
2. Eggs	2
3. Onion medium	1
4. Ginger	1 inch piece
5. Garlic	2 pods
6. Red chilli powder	½ tsp
7. Vinegar	1 tbsp

8. Bread crumb	as needed
9. Salt	to taste

Method

1. Take jumbo prawns without head, wash them and remove the shell, devein. Grind ginger, onion and garlic together adding little water.
2. Marinate the prawn in the juice of onion, ginger and garlic and add vinegar, red chilli powder, and salt to it, set aside for thirty minutes.
3. Separate the white of the eggs and beat properly.
4. Dip prawns one by one in it and then roll in the bread crumb.
5. Heat oil very hot in the pan and add one by one the prawns and deep fry.
6. Reduce the flame and fry until light brown.

P11) Prawn Kabab Recipe

Prawn Kabab is a delicious Malabar recipe. Learn how to make/prepare Prawn Kabab by following this easy recipe.

Ingredients

1. Prawns	10
2. Garlic paste	2 tsp
3. Ginger paste	2 tsp
4. Lemon juice	4 tsp
5. Curd	2 cups
7. Cream	2 cups
8. Gram flour	4 tsp
9. White flour	2 tsp

10.	Cumin seeds	2 tsp
11.	Clove powder	2 tsp
12.	Salt	to taste

Method

1. Clean the prawns.
2. Shed them for 10 minutes.
3. Blend the curd and cream properly.
4. Add gram flour, white pepper and clove powder.
5. Mix it well with prawns.
6. Add salt and put the mixture in silver foil.
7. Keep the prawns in refrigerator for 2 hours.
8. Put prawns in an oven and tandoor it for 20 minutes.
9. Once the prawns turn golden brown, put butter over it. Oven them for 2 minutes.
10. Prawn Kabab is ready.

P12) Prawn Pulav Recipe

Prawn Pulav is a delightful recipe. Learn how to make/ prepare Prawn Pulav by following this easy recipe.

Ingredients

1.	Prawns (Chopped)	3 cups
2.	Rice	3 cups
3.	Turmeric powder	1 tsp
4.	Tandoori masala	1 tsp
5.	Garlic flakes	2
6.	Ginger	1
7.	Chillies	4

8.	Onion	1
9.	Cloves	4
10.	Peppercorns	4
11.	Cardamoms	2
12.	Coriander Chopped	2 tbsp
13.	Water	6 cups
14.	Oil	for frying
15.	Salt	to taste

Method

1. Wash and clean the prawns.
2. Wash and clean the rice. Grind garlic, ginger and chillies to make a paste. Chop the onions.
3. Add chilli powder, turmeric powder, salt and tandoori masala to the cleaned prawns.
4. Mix it well. Keep the mixture aside. Heat the oil.
5. Add cloves, peppercorns and cardamoms.
6. Mix garlic ginger chilli paste. Fry it and chopped onions.
7. Fry the onions till translucent. Add washed rice and fry for few minutes.
8. Now add prawn mixture and fry for few minutes.
9. Pour 6 cups of water and cook till rice is properly cooked.
10. Adorn with chopped coriander leaves.

P13) Prawns Koliwada Recipe

Prawns Koliwada is a yummy Punjabi recipe. Learn how to make/prepare Prawns Koliwada by following this easy recipe.

Ingredients

1. Prawns	400 gm Shelled and deveined
2. Plain flour	1 tbsp
3. Garlic paste	1 tsp
4. Red chilli powder	2 tsp
5. Ginger paste	1 tsp
6. Garam masala	1 tsp
7. Thymol seeds	½ tsp
8. Lemon juice	1 tbsp
9. Oil	for deep frying
10. Salt	to taste

A pinch of orange–red colour

Onion slice and lemon wedges to garnish.

Method

1. Wash the shelled and deveined prawns properly.
2. In a bowl, combine the prawns with the remaining ingredients until the prawns are well-coated with the spices. Marinate for 3 hours.
3. Heat the oil until it is very hot and deep fry the prawns over medium flame until they are crisp.

4. Garnish it with onion slices and lemon wedges and serve.

P14) Spicy Prawn Recipe

Spicy Prawn is a piquant Malabar recipe. Learn how to make/prepare Spicy Prawn by following this easy recipe.

Ingredients

1.	Prawns	500 gm
2.	Onions	3
3.	Ginger	1
4.	Chillies	4
5.	Garlic flakes	4
6.	Coriander seeds	4 tbsp
7.	Poppy seeds	2 tbsp
8.	Grated coconut	3 tbsp
9.	Cinnamons	4
10.	Cloves	6
11.	Cardamoms	4
12.	Oil	for frying.

Method

1. Chop onions into slices. Grind coriander, poppy seeds, grated coconut, garlic flakes, cinnamons, cloves and cardamoms to make a paste.
2. Grind ginger and garlic to make a paste.
3. Heat the oil. Fry onion slices and chillies till brown.
4. Add ginger garlic paste.
5. Now add prawns, turmeric powder and salt.
6. Fry for 10 minutes. Mix chilli powder and prepared paste.

7. Stir well. Add water and cook till the mixture turns in gravy.
8. Spicy Prawn is ready.

P15) Choo Chee Prawn Recipe

Choo Chee Prawns is a delicious recipe. Learn how to make/prepare Choo Chee Prawns by following this easy recipe.

Ingredients

1. Prawns	1 cup
2. Mushrooms	½ cup
3. Coconut cream	2 cups
4. Jalapeno peppers	3
5. Dry red chillies	2
6. Garlic flakes	12
7. Shrimp paste	2 tsp
8. Fish sauce	4 tbsp
9. Shallots	6 chopped

10.	Peanut oil	2 tbsp
11.	Lime leaves	4 tsp

Fresh coriander leaves for garnishing.

Method

1. Grind red peppers and shrimp paste to make a well blended paste.
2. Shell the prawns. Add salt and 2 tbsp water.
3. Mix it well and keep it aside for 1 hour. Rinse with running water.
4. Drain all the water. Heat the oil. Fry the mushrooms till the colour is changed.
5. Add fish sauce and cook till tender. Keep it aside.
6. Heat the pan. Put 1 cup of coconut cream and heat till cream is thickened.
7. Stir the prepared paste till it turns aromatic.
8. As oil is separated from cream, season it with fish sauce and palm sugar.
9. Add prawns and lime leaves once the mixture turns thick.
10. Fry till prawns are cooked. Toss the cooked mushrooms with prawns.
11. Reheat and mix in chilli sauce. Choo Chee Prawns is ready.

P16) Prawn Fajitas Recipe

Prawn Fajitas is a delicious Mexican recipe. Learn how to make/prepare Prawn Fajita by following this easy recipe.

Ingredients

1.	Green King Prawns medium	200gm Peeled
2.	Tortillas (flour)	8
3.	Fresh lime juice	1 tbsp
4.	Sour cream	½ cup
5.	Fresh coriander	1 ½ tbsp coarsely chopped
6.	Olive Oil	1 tbsp
7.	Red onion sliced	1
8.	Green capsicum	½ deseeded, thinly sliced
9.	Tomato	1 large finely chopped
10.	Red capsicum	½ deseeded finely chopped
11.	Ground cumin	½ tsp
12.	Mild paprika	½ tsp
13.	Ground coriander	½ tsp
14.	Large avocado	1 peeled and chopped

Method

1. Preheat the oven to 180 degrees C.
2. Wrap the tortillas in foil and put it in the preheated oven for about 10 minutes till heated through.
3. Remove it from the oven.
4. Combine the sour cream, lime juice and fresh coriander in a small bowl, mix well.
5. Meanwhile, heat half the oil in a large non-stick frying pan on medium-high flame.
6. Add the onion, red and green capsicum and cook, for about 2-3 minutes till onion softens, keep stirring. Place it to a bowl.

7. Take a pan and heat rest of the oil in the pan over high flame.
8. Add the prawns and cook, for about 1-2 minutes till prawns curl and change the colour.
9. Combine the onion mixture, tomato, cumin, ground coriander and paprika to the prawns and cook for about 1-2 minutes till aromatic. Remove it from the fire.
10. Divide the prawn mixture equally among the tortillas.
11. Top it with the avocado and a dollop of sour-cream mixture.
12. Now fold in half or roll up the tortillas and serve it.

P17) Prawn Risotto Recipe

Seafood Risotto is a mouth watering recipe, perfect for any occasion. Learn how to make/prepare Seafood Risotto by following this easy recipe.

Ingredients

1. Fresh prawns 200 gm
2. Squid fresh 100 gm

3.	Mussels	100 gm
4.	Arborio rice	60 gm
5.	White wine	75 ml
6.	Oil	40 ml
7.	Fresh shallots, Chopped	10 gm
8.	Butter	20 gm
9.	Saffron	1/16 gm
10.	Cheese	30 gm
11.	Garlic, chopped	2 gm

Method

1. Fry the prawns, mussels and squid, in oil.
2. Combine chopped shallots and chopped garlic with it.
3. Add the rice to it and cook for a while.
4. Now add the white wine and saffron and cook for about 10 minutes.
5. Add the butter to it and keep stirring.
6. Add the cheese and serve Seafood Risotto.

P18) Prawn Tikka Recipe

Prawn Tikka is a Malabar appetizer recipe. Learn how to make/prepare Prawn Tikka by following this easy recipe.

Ingredients

1.	Prawns Cleaned	300 gm
2.	Chilli powder	1 tsp
3.	Garlic flakes	4
4.	Lemon flakes	4 tsp
5.	Sesame seeds	3 tbsp

6. Coriander leaves	1 bunch
7. Salt	to taste
8. Oil	for frying.

Method

1. Cut prawns into pieces.
2. Grind garlic to make a paste.
3. Make a spot horizontally with a knife in the middle.
4. Mix lemon juice, salt and garlic paste.
5. Put this mixture in the middle of the prawn pieces.
6. Keep prawns for drying.
7. Grind sesame seeds, chilli powder, salt and coriander leaves to make a paste.
8. Put this paste on the prawn pieces.
9. Fry these prawn pieces till crisp. Serve hot.

P19) Sungta Humman Recipe

Sungta Humman is a very popular recipe. Learn how to make/prepare Prawn Curry by following this easy recipe.

Ingredients

1.	Headless prawns shelled and cleaned	500 gm
2.	Coconut oil	2 tbsp
3.	Asafetida powder	1 tsp
4.	Coconut grated	1
5.	Red chillies roasted in coconut oil	8
6.	Tamarind paste	1 tbsp
7.	Salt	to taste

Method

1. Grind grated coconut, chillies, and tamarind paste taking water as required to make a fine paste.
2. Mix the ground paste, prawns, asafetida powder and salt in a pan.
3. Add water and mix well. Bring it to boil on low heat.
4. Add in the coconut oil and keep on medium heat till the prawns are cooked and the gravy is thick.

P20) Kolmino Patio Recipe

Kolmino Patio is a tempting Parsi recipe. Learn how to make/prepare Kolmino Patio by following this easy recipe.

Ingredients

1.	Prawns, cleaned	1 cup
2.	Garlic flakes	6
3.	Onion sliced	1
4.	Onions Chopped	3

5.	Turmeric powder	2tsp
6.	Cumin seeds	¼ tsp
7.	Tamarind juice	½ cup
8.	Red chillies	5
9.	Jaggery grated	1 tsp
10.	Oil	4 tsp
11.	Salt	to taste
12.	Fresh coriander leaves	

Method

1. Grind garlic, turmeric powder, cumin seeds and red chillies to make a paste.
2. Heat the oil and fry sliced onions till golden brown.
3. Add prepared paste.
4. Cook it for 5 minutes.
5. Mix chopped onions, coriander leaves and salt.
6. Add water and allow prawns to get cooked.
7. Mix tamarind juice and jaggery. Cook till jaggery is melted.
8. Adorn with fresh coriander leaves.

P21) Schezuan Fried Rice Recipe

Schezuan Fried Rice is a palate-teasing blend of spices. Learn how to make/prepare Szechuan Fried Rice by following this easy recipe.

Ingredients

1.	Prawns shredded	1 cup
2.	Chicken shredded	1 cup
3.	Basmati rice	250 gm

4.	Green capsicum medium	1
5.	Spring onion, chopped	3
6.	Small shrimp, cooked	
7.	Chilli sauce	½ tbsp
8.	Carrot, cleaned, chopped	1
9.	French beans finely chopped	¼ cup
10.	Celery, cleaned chopped	1 tbsp
11.	Vinegar	1 tbsp
12.	Worcestershire sauce	1 tbsp
13.	Ajinomoto	½ tbsp
14.	Onion sliced	1
15.	Oil	5 drops
16.	Garlic, chopped	1
17.	Tinned mushrooms, chopped	½ cup
18.	Salt	to taste

Method

1. Take a vessel and pour one tsp of oil.
2. Fry the onion until transparent.
3. Add, capsicum, celery, spring onion, carrots, beans and mushrooms.

4. Cook them for a few minutes until they turn brown.
5. Take another vessel and fry the chicken and prawns.
6. Pour a little oil. Add chopped garlic.
7. Remove it when it turns brown. Keep aside.
8. Once oil is flavoured with garlic, place the rice in the same vessel.
9. Now Fry it for 5 minutes.
10. Combine all the vegetables, chicken and prawns.
11. Cook the rice in the chicken stock.
12. Mix all the sauces with cooked rice.
13. Fry for 5 minutes on high heat.
14. Keep stirring and serve hot.

P22) Spicy Prawn Salad

Ingredients

1. Prawns	12
2. Tomato Chopped	1
3. Parsley finely chopped	2tbsp
4. Mix together for dressing:	
5. Lemon grass finely chopped	2 tbsp
6. Any sauce	3tsp
7. Green chillies finely chopped	5
8. Onion small finely chopped	1
9. Mint leaves crushed	1tsp
10. Small raw mango chopped	1
11. Vinegar	2 tbsp
12. Lemon juice	1 tbsp

13.	Olive oil	1 tbsp
14.	Salt	to taste.

Method

1. Refrigerate the dressing for 1 hour.
2. Clean and devein prawn. Slit them keeping intact.
3. Boil in hot water till cooked.
4. Drain and keep aside till cool. In a salad dish, mix prawn and tomato.
5. Pour the dressing over it. Toss the salad.
6. Garnish with parsley leaves before serving.

P23) Tangi Prawn and Kiwi Salad (Chinese Recipe)

Ingredients

1.	Cooked prawns	250 gm
2.	Kiwi peeled and sliced	1
3.	Orange halves shells only	2
4.	Oranges, segmented	2
5.	Nuts chopped	2 tbsp
6.	Herbs chopped	4 tbsp

For Dressing

1.	Yoghurt	2 tbsp
2.	Lemon juice	2 tbsp
3.	Honey	1 tbsp
4.	Mustard	½ tsp

Method

1. Combine orange segments with prawns, kiwi, nuts and herbs.

2. Season it Whisk together dressing ingredients.
3. Mix with prawns and fruit.
4. Fill in to shells. Chill in refrigerator.
5. Flatten bread slices with a rolling pin. Beat butter, rind, herbs and pepper together.
6. Spread over bread and scatter mushrooms.

P24) Prawn Sushi (Butterfly Style)

Cooking Time: 15-30 mins

Ingredients

1. Prawns, Unshelled	6

For Filling

1. Egg yolks	4
2. Sugar	1 ¼ tsp
3. Lemon juice	2 ½ tsp
4. Salt	to taste

For Marinade

1. Rice vinegar	3 tbsp
2. Sugar	1 tsp
3. Ajinomto	a pinch
4. Water	4 tbsp
5. Salt	¼ tsp

Method

1. Keeping the shell intact pass a tooth pick through the flesh of the prawn so that it doesn't curl.
2. Boil these prawns in water for about 3–4 minutes or till they change colour.

3. Drain the water and immediately pass the prawns through cold water.
4. Now remove the toothpicks and peel the shell off from the prawns, but leaves the tail section of the shell.
5. Cut the prawn slightly and remove the vein from its back.
6. Make a deeper incersion on the prawn and spread the prawn making it look like butterfly.
7. To prepare the marinade, mix the vinegar with water, salt, sugar, Ajinomoto and mix it with the prawns, then marinate for 1 hour.
8. Beat the egg yolks lightly and remove about 2 ½ tsp of the egg yolks and keep aside.
9. Add water to the remaining yolks, mix well and cook over a medium flame till the yolks turn firm.
10. Drain the yolks and then mash them to a fine paste.
11. Mix the sugar, salt, lemon juice, ajinomoto and the rest of egg yolk and beat well.
12. Pass this mixture through a fine sieve, then mix it with the mashed egg yolk and keep aside.
13. Divide the filling into 6 portions and then filling them into the 6 individual prawns.
14. Seal the edges but pressing them together and serve.

P25) Prawn Dalna

Preparation time: 15–30 mins

Ingredients

1. Prawns 250 gm

2.	Onions	3 medium
3.	Potatoes large	3-4
4.	Haldi powder	1tsp
5.	Mirchi powder	¾ tsp
6.	Green chillies	4–5 according to taste
7.	Coconut powder	50 gm
8.	Fresh cream	According to taste

Method

1. Wash prawns (or shrimps) and marinade with coconut powder (3-4 teaspoons), Haldi Powder (½ teaspoon), Salt (1 teaspoon) and fresh cream. Leave for 15-20 mins.
2. Boil the potatoes till done. It is important that they are not over-done. The pressure-time (or boiling time) varies with the potatoes type, use own judgment to assure that they are just cooked but firm.
3. Peel off skin and cut in triangular (not too small) pieces.
4. Heat Oil, add above potatoes, chillies, onions, haldi powder (¼ teaspoon) and coconut powder (3-4 teaspoons). Salt according to taste.
5. Fry till the onions and potatoes are brown.
6. Add fresh cream. Stir 1 min.
7. Add marinated prawns along with the marinade. Mix well.
8. Put the entire mix in an Oven and 280 degrees Celsius preheated for about 3-5 minutes. For the

prawns, I prefer a harder crust and soft inside, you may use your discretion.

9. In case there is no oven available, heat on a medium flame for 7-8 minutes.

This is best eaten with Basmati Rice.

P26) Prawn Toast

Preparation time: 45-60 mins

Ingredients

1.	Prawn meat	2 cups
2.	Slices of bread, sides removed	2
3.	Sesame seeds oil for deep frying (Grind together for smooth mixture)	½ cup
4.	Onion, chopped finely	1
5.	Egg white, beaten	1
6.	Pepper	as per taste
7.	Coriander leaves (chopped)	2 tsp
8.	Garlic powder	¼ tsp
9.	Bamboo shoots	2 tsp
10.	Water	2 tbsp

Method

1. Slice the breads into 4 triangles
2. Spread mixture evenly on the breads
3. Put the sesame seeds in a flat plate
4. Put the mixture coated side of bread into sesame seed plate
5. Press a lit, so the slice is completely coated with sesame seeds

6. Heat oil
7. Deep fry till light golden, from both sides
8. Remove and drain
9. Serve hot

P27) Prawn Balls

Preparation time: 15-30 mins

Ingredients

1.	Cleaned prawns	250 gm
2.	Oil	2 tbsp
3.	Finely chopped onion	2 tbsp
4.	Finely chopped garlic	1 tsp
5.	Coriander powder	½ tsp
6.	Chilli powder	1 tsp
7.	Turmeric powder	½ tsp
8.	Salt & pepper	to taste
9.	Bread	4 slices
10.	Milk	1/3 cup
11.	Eggs	2, lightly beaten
12.	Egg and bread crumbs to coat	

Method

1. Saute the prawns with onion and garlic in oil. Add coriander and chilli, turmeric powder, salt and pepper.
2. When the prawns are cooked, cool and mince the mixture.
3. Remove crust from bread, cut into pieces and soak in milk. Mash and add to prawn mixture.

4. Add lightly beaten eggs and mix well. Cook the mixture on low flame, stirring all the time till it is thick enough to hold its shape.
5. Cool and shape into bite size balls. Roll in bread crumbs, then in beaten egg and in bread crumbs again. Deep fry till golden brown. Serve hot.

P28) Prawn Cutlets

Preparation Time: 15–30 mins

Ingredients

1.	Prawn	½ Kgs
2.	Onion	1
3.	Ginger	1 inch piece
4.	Curd	100 gm
5.	Red Chillies powder	as per taste
6.	Salt	to taste
7.	Haldi powder	½ tsp
8.	Garam Masala powder	1 tsp
9.	Egg	
10.	Bread crumb	1 cup
11.	Oil	for frying

Method

1. Fork the prawns.
2. Flatten the slitted sides.
3. Make the paste of onion mix in it garam masala, curd, salt haldi, chilli.
4. Apply this paste on prawns and keep them aside for about half an hour.

5. Beat the eggs mix salt in it
6. Heat the oil
7. Dip the prawns in egg and apply bread crumb and fry it .

Prawn cutlets are ready to serve (with sauce or salad)

P29) Prawn Surprise

Preparation Time: 15-30 mins

Ingredients

1.	Prawns (shelled)	½ kg
2.	Mayonnaise	4 tbsp
3.	Onion (Chopped)	2 tbsp
4.	Grated cheese	2 tbsp
5.	Thick cream	1 tbsp
6.	Salt	to taste
7.	Pepper and mustard	to taste
8.	Peppercorns	Few
9.	Green chillies (Sliced)	
10.	Slice of bread	6

Method

1. Boil the shelled prawns with salt, peppercorns, sliced green chillies and very little water (about ¼ cup).
2. When prawns turn pink and are cooked, drain and cool, removing peppercorns and chillies. Chill the prawns.
3. Combine the boiled prawns with mayonnaise, chopped spring onions, grated cheese, thick cream,

salt, pepper and mustard to taste (if the mixture looks dry, add more cream).

4. Cut the bread slices into squares. Pile prawn mixture on each square. Sprinkle a little more grated cheese on top.
5. Bake in a very hot oven till squares are golden-brown and bubbly.

Serve hot decorated with parsley.

P30) Prawn Baffad

Preparation Time: 30-45 mins

Ingredients

1.	Prawns	250 gms
2.	Onion Chopped	2
3.	Big tomato skinned and chopped	2
4.	Kashmiri chillies dried	8–10
5.	Cummin seeds	½ tsp
6.	Peppercorns	12
7.	Garlic cloves	3
8.	Ginger	½ inch
9.	Turmeric powder	1 tsp
10.	Salt	to taste
11.	Vinegar	to taste

Method

1. Grind together the chillies, cummin, peppercorns, garlic and ginger.
2. Mix in the turmeric in half cup of water.
3. Fry the onions till light brown.

4. Add the tomato and stir a bit before adding the prawns.
5. Fry for a minute.
6. Add the masala, tamarind and vinegar.
7. Cover and cook until prawns are just done. Do not overcook.

Serve hot

P31) Prawn Vindaloo

Preparation Time: 15-30 mins

Ingredients

1. Prawns	1 kg
2. Onion	2 medium size
3. Cumin seeds	1 tbsp
4. Coriander seeds	1 tbsp
5. Mustard seeds	1 ½ tbsp
6. Vinegar	2 tbsp
7. Turmeric powder	½ tbsp
8. Red chilli powder	1 tbsp
9. Mustard oil	1 tbsp
10. Salt	to taste

Method

1. Peel and wash the prawns and keep aside.
2. Grind cumin, coriander and mustard seeds together. Add the grinded spices, oil, chilli powder, salt, turmeric powder, vinegar into the prawns and mix it well and keep aside for an hour.
3. In a wok add 2 tbsp of mustard oil and fry the onions and then add prawns.

4. Cook it in a low heat and cover it till it is cooked. Have it with steamed rice.

P32) Fried Prawn with Corn Flour

Ingredients

1.	Fresh prawns	250 gm
2.	Maida	½ cup
3.	Corn flour	½ cup
4.	Aginomotto	1/8 tsp
5.	White of one egg	
6.	Powdered salt	½ tsp
7.	Baking powder	Three pinches

Method

1. Mix Maida, corn flour, aginomotto, salt and baking powder.
2. Beat the egg white thoroughly and mix in the above powder.
3. Dip each prawn in the batter and fry in oil.
4. Serve hot.

P33) Prawn Chilli Fry

Ingredients

1.	Prawns (Shelled)	250 gm
2.	Onions (Sliced)	6
3.	Green chillies	2–3
4.	Ginger	½ " piece
5.	Red Chilli powder	¼ tsp
6.	Turmeric	½ tsp

7. Ground pepper — ¼ tsp
8. Salt — to taste.

Method

1. Fry the onion till soft.
2. Add the green chillies, ginger and powdered masalas.
3. Lastly add the prawns and cook till dry.

This should be eaten together with rice and curry.

Serves–4

P34) Fried Prawns

Ingredients

1. Prawns (shelled) — 250 gm
2. Turmeric — 1 tsp
3. Chilli powder — ½ tsp
4. Lemon Juice — ½ lemon
5. Salt — to taste.

Method

1. Marinate the prawns in a turmeric, chilli powder, salt and lemon juice.
2. Keep it for half an hour.
3. Fry till brown.
4. Serve hot.

P35) Jheenga Biryani

Ingredients

1. Medium size prawns — 24
2. Desi Ghee — 60 gm

3.	Cardamom	3
4.	Cloves	2
5.	Cinnamon	2–5
6.	Bay leaves	2
7.	Ginger–garlic paste	40 gm
8.	Curd	60 gm
9.	Chilli powder	2 gm
10.	(a) Ginger julienne	300 gm
	(b) Green chilli (chopped)	3
	(c) Mint leaves (chopped)	10 gm
	(d) Coriander leaves	15 gm
	(e) Onion sliced	25 gm.
11.	Prawn stock/water	350 ml
12.	Lemon juice	20 ml
13.	Cream	60 ml.
14.	Saffron	few strands in little milk.

Method

To Prepare Prawns

1. Clean, devein and shell the prawns, but retain the tail.
2. Reserve the shells for preparing prawn stock..
3. Colour the shells in an oven and then add water and simmer to extract the flavour.
4. Heat ghee in Kadai,
5. Add the whole garam masala, stir for a few seconds.
6. Add ginger garlic paste and keep stirring for three minutes.

7. Lower the heat and add curd, chilli powder and salt.
8. Stir–fry for five minutes.
9. Add prawns and stir-fry for five minutes.
10. Add the stock and cook till the prawns are done.
11. Add lemon juice, cream cardamom and adjust the seasoning.

To Assemble the Biryani

1. Bring plenty of water to a boil.
2. Add salt and rice and cook till the rice is three-fourth cooked.
3. Drain and keep aside.
4. Remove half of the gravy from the prawn mixture and returns kadai to heat.
5. Spread half of the ginger juliennes, mint leaves and coriander leaves.
6. Spread half of the partially cooked rice. Repeat the layers.
7. Sprinkle fried onions saffron milk and lime juice .
8. Cover with lid, seal with atta dough and cook on low heat, or pre-heated oven for 10–15 minutes until the steam begins to seep out of the dough.
9. Serve hot.

P36) Creamy Shrimp and Veggies

Ingredients

1.	Shrimp cooked and peeled	500 gm
2.	Red pepper	1 tsp

3.	Onion chopped	100 gm
4.	Mushroom frozen	100 gm
5.	Butter	1 tbsp
6.	Cream	1/3 cup
7.	Garlic	1 tsp
8.	Parmesan cheese	1 tbsp

Method

1. Saute Red pepper and mushrooms in Butter until warm.
2. Add shrimp, cream, and garlic.
3. Add chopped onion.
4. Simmer on low to medium heat until veggies are done and mix is heated thoroughly.
5. Top with Parmesan cheese and serve.

P37) Shrimp Rosti

Ingredients

Rosti

1.	Small shrimps peeled and deveined	250 gm
2.	Potatoes, peeled and finely grated	325 gm
3.	Carrot peeled and finely grated	1
4.	Onion, finely grated	1 small
5.	Flour	4 tbsp
6.	Egg	1
7.	Pepper	to taste

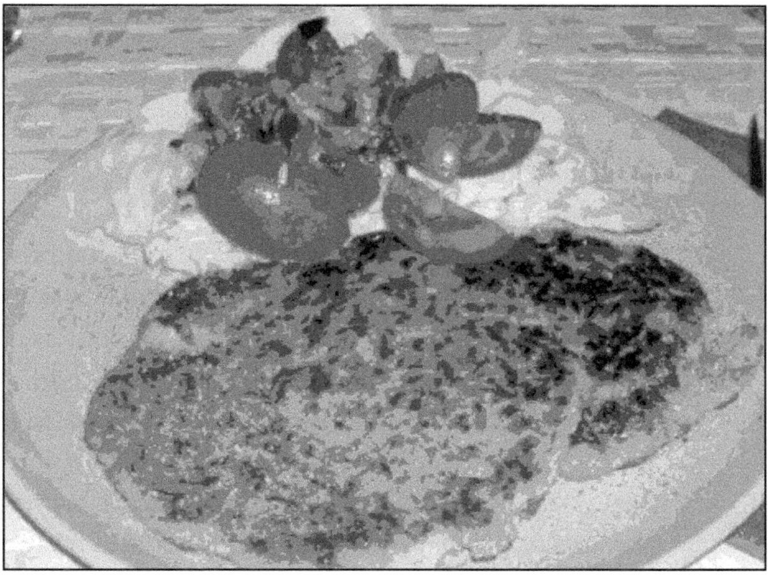

8.	Salt	to taste
9.	Oil	to fry.

Tomato-Salsa

1.	Tomato quartered	250 gm
2.	Mango diced	½
3.	Chilli, finely chopped	1
4.	Red Onion finely chopped	1 small
5.	Fresh cilantro chopped	1 tbsp
6.	Fresh green onion	1 tbsp
7.	Olive oil	2 tbsp
8.	Lemon juice	2 tbsp
9.	Pepper	to taste
10	Salt	to taste.

Method

Tomato Salsa

1. In a bowl, mix all the ingredients together. Taste with salt and pepper. Reserve.

Rosti

1. Add in a bowl the potatoes, celery, shrimps, onions, flour and egg. Mix well all the ingredients. Season with salt and pepper.
2. Heat some oil in a non stick frying pan, add a portion of the mixture and spread with a spatula so that the mixture covers the pan. Cook on a low heat pressing occasionally, until the rosti is brown and crispy on the underside. Flip over and let the other side cook. Repeat the process until the mixture is finished.
3. Serve immediately with the tomato-salsa.

P38) Shrimp with Garlic

Ingredients

1.	Unshelled shrimp	450 gm
2.	Garlic, finely chopped	4 cloves
3.	Chillies, chopped	2
4.	Sparsely chopped	4 tbsp
5.	Pepper	to taste
6.	Salt	to taste.

Method

1. Heat olive oil in a large frying pan, stir in garlic and chilies peppers for 1 to 2 minutes.

2. Add shrimps, stirring for 3 to 5 minutes or until they turn red.
3. Remove frying pan from heat, sprinkle fresh parsley.
4. Season with salt and pepper.
5. Transfer shrimps to a serving bowl.
6. Garnish with lemon slices

P39) Dry Prawns Peera

Ingredients

1. Dry prawns cleaned, soaked in water and drained	1 cup
2. Grated coconut	1 cup
3. Raw Mango	¼
4. Green chilli	1
5. Curry leaves	1 stem
6. Turmeric powder	½ tsp
7. Salt	to taste.

Method

1. Grind together the grated coconut and turmeric powder.
2. Add the paste to a kadai along with chopped green chilli, raw mango, curry leaves, water and salt.
3. Heat the kadai and when the gravy starts to boil, add the dry prawns.
4. Allow to cook until the gravy thickens.

P40) Prawn Fry

Ingredients

1. Prawns	250 gm
2. Onion chopped	1 cup
3. Tomato chopped	¾ cup
4. Ginger	1 inch
5. Garlic	7 cloves
6. Turmeric powder	¾ tsp
7. Red chilli powder	1 ½ tsp
8. Aniseed	¾ tsp
9. Curry leaves	few
10. Oil	3 tsp
11. Salt	to taste.

Method

1. Clean the prawns. Grind the ginger, garlic and aniseed to a fine paste.
2. In a heavy bottom Kadai, add oil and heat it.
3. Add the ginger-garlic paste and fry for a minute.

4. Add tomato, onion, chilli powder, turmeric powder and salt.
5. Fry this for 4 minutes.
6. Add the prawns and cook till the prawns become tender.
7. Then add curry leaves and fry it well.

P41) Prawns in Pink Sauce

Ingredients

1.	Peeled prawns	900 gm
2.	Garlic, chopped	3cloves
3.	Mustard seeds	1 tbsp
4.	Oil	3 tbsp
5.	Cumin ground	1 tsp
6.	Pepper	1 tsp
7.	Lemon juice	20 ml
8.	Cayenne	½ tsp
9.	Ginger grated	1 tsp
10.	Cream	1 cup
11.	Tomato puree	¾ cup
12.	Sugar	½ tsp
13.	Salt	to taste

Method

1. Mix together the tomato puree, water, ginger, cayenne, lemon juice, salt, pepper and sugar.
2. Heat the oil and fry the mustard seeds until they start crackling.
3. Add the garlic then the prawns and stir-fry for 2 minutes.

4. Pour in the sauce mixture and stir over a low heat until the sauce starts bubbling.
5. Serve at once with rice and vegetables.

P42) Tangi Prawns with Mushroom Rolls

Ingredients

1.	Cooked Prawns	250 gm
2.	Kiwi peeled and sliced	1
3.	Orange halves Shells only	2
4.	Oranges segments	2
5.	Chopped Nuts	2 tbsp
6.	Chopped herbs	4 tbsp
7.	Oil	1 tbsp
8.	Pepper	to taste
9.	Salt	to taste

For Dressing

1.	Yoghurt	2 tbsp
2.	Lemon juice	2 tbsp
3.	Honey	1 tbsp
4.	Mustard	½ tsp

For Mushroom Rolls

1.	Bread, crusts removed	4 slices
2.	Softened butter	1 tbsp
3.	Lemon rind	¼ tsp
4.	Chopped herbs	1 tsp
5.	Finely chopped Mushroom, fried	2 tbsp
6.	Pepper	to taste.

Method

1. Combine orange segments with prawns, kiwi, nuts and herbs. Season it.
2. Whisk together dressing ingredients.
3. Mix with prawns and fruit.
4. Spoon into orange shells.
5. Chill
6. Flatten bread slices with a rolling pin.
7. Beat butter, rind, herbs and pepper together.
8. Spread over bread and scatter mushrooms
9. Roll up bread. Chill well before serving

P43) Prawn Kofta

Ingredients

1. Peeled prawns	450 gm
2. Onion, finely chopped	3
3. Green chillies finely chopped	2
4. Egg, beaten	1
5. Bread crumbs	½ cup
6. Mustard oil	2 tbsp
7. Ghee	2 tbsp
8. Bay leaves	2
9. Garam masala	2 tsp
10. Turmeric ground	2 tsp
11. Ginger fresh chopped	2 tsp
12. Coconut milk	1 ¼ cup
13. Coriander leaves chopped	1 tbsp

Method

1. Season the prawns with salt then mix with the onions and chilies.
2. Shape into 12 balls or koftas, dip in egg the roll in breadcrumbs.
3. Heat the mustard oil and fry until browned.
4. Heat the ghee and fry the bay leaves, onion, garam masala, turmeric and ginger for 5 minutes, stirring continuously and adding a little water as necessary.
5. Stir in the coconut milk and salt and bring to the boil.
6. Add the koftas and simmer for 10 minutes until the sauce is thick.
7. Garnish with coriander and serve hot with rice.

Chapter 3
Other Seafood Recipes

Lobster Recipe

Lobsters are probably the most sophisticated shellfish; a lobster recipe is a dish that stands out on any dinner table. With a lobster dish you can impress your family or guests. You can prepare lobster recipes using different cooking methods, from the simplest one - boiling them - until grilling

or baking them. Learn how to cook lobster with step-by-step instructions.

Lobsters are also delicious and very nutritious - it contains less cholesterol, calories, and saturated fats than lean beef, pork, and chicken. Just choose below the lobster recipe you like most and enjoy this delicious and sophisticated seafood.

L1) Lobster Pappas

Ingredients

1. Cleaned lobster	500 gms
2. (a) Coriander powder	1 desert spoon
(b) Chilli powder	½ tsp
(c) Turmeric powder	¼ tsp
(d) Pepper powder	1/8 tsp
(e) Fenugreek	1/8 tsp
(f) Cloves of garlic	2
3. Oil	2 desert spoon
4. Mustard	¼ tsp
5. (a) Finely chopped onion	½ cup
(b) Small cloves of garlic	1 tsp.
(c) Ginger cut lengthwise	½ tsp
6. (a) Water	1 ½ tsp
(b) Malabar tamarind water	¼ cup
7. (a) Split green chillies	2
(b) Curry leaves	few

Method

1. Finely grind the second ingredients into masala.

2. Fry the mustard in oil, till the mustard bursts and add the fifth ingredients in the order, mentioned and fry till brown.
3. To this add the finely ground masala.
4. Mix the sixth ingredients and allow it to boil.
5. Then add the seventh ingredients and stir till the gravy thickens.
6. Serve hot.

L2) Lobster with Cashew

Ingredients

1.	Lobster	1 cup
2.	Cashewnut	1/8 cup
3.	Coconut oil	1/8 cup
4.	Sliced onion	¼ cup
5.	Ground ginger	½ tsp
6.	Ground garlic	¼ tsp
7.	Tomato	1
8.	Chilli powder	¼ tsp
9.	Pepper powder	¾ tsp
10.	Turmeric powder	¼ tsp
11.	Green chilli split	4
12.	Vinegar	1 tsp
13.	Salt	To taste
14.	Ground semi fried coconut scrapping	1 spoon

Method

1. Split the cashewnuts and soak in hot water.

2. Saute the onion, ginger, garlic and tomato in coconut oil one by one.
3. Add the lobster and soaked cashewnut, followed by ingredients 8,9, and 10.
4. Stir continuously so that it does not stick to the bottom of the pan.
5. Add sufficient water to cook the lobster meat with salt.
6. When the water reduces to ¾ add green chillies and vinegar and cook till gravy starts thickening.
7. To it add the 14th ingredient mixed in ¼ cup of water.
8. Cook for 10 minutes till the gravy thickens.

L3) Lobster Stuff Potatoes

Potatoes stuffed with a lobster meat, cheese and sour cream mixture - this is not exactly a light meal, but it does not matter because it is simply delicious!!

Ingredients

1. Fresh lobster meat	225 gm
2. Fresh mushrooms	½ cup
3. Potatoes	6
4. Onion	1 large
5. Garlic diced	2 cloves
6. Scallions, finely chopped	2
7. Sour cream	1 tsp
8. Fresh parmesan cheese	½ cup
9. Gruyere cheese	½ cup
10. Vermouth	¼ cup

11. Black pepper fresh ground ½ tsp
12. Egg beaten well 1

Method

1. Bake potatoes for 45 minutes in 400 degree oven until tender. Scoop out inside of potatoes, leaving ¼ inch of potato in skin. Set aside.
2. In skillet, Sauté onion and garlic until transparent, in about 2 tablespoons butter. Add mushrooms, Sauté 3-5 minutes.
3. Meanwhile chop lobster meat into bite-size pieces.
4. Add to sauté mixture, Sautee about 3 minutes, until it's heated through, turn burner to medium high, add the vermouth, cook until vermouth boils down. Set aside.
5. Meanwhile, add sour cream to potatoes and hand mash together.
6. Add the cheddar and Gruyere cheese, mix together.
7. Add plenty of fresh-ground pepper, beaten egg and lobster mixture; blend together.
8. Add scallions and ¼ cup of Parmesan.
9. Stuff potatoes with mixture, sprinkle with remaining Parmesan cheese.
10. Bake at 350 degrees until golden brown (about 15 minutes).

L4) Lobster with Coconut Milk

Coconut milk is an ingredient that combines very well with different kinds of seafood. Try this recipe prepared with lobster meat and coconut milk. This dish is to be served with rice.

Ingredients

1.	Lobster meat cut into cite size pieces	900 gm
2.	Butter	1 tbsp
3.	Flour	1 tbsp
4.	Dry white wine	4 tbsp
5.	Ketchup	½ tbsp
6.	Onion finely chopped	1 small
7.	Whipping cream	½ cup
8.	Pepper	1 tsp
9.	Salt	to taste
10.	Tomatoes	2

Method

1. In a bowl, put the lobster meat and the wine. Season with salt and pepper. Set aside.
2. In a saucepan, melt the butter and saute the onion.
3. Tomatoes to be blanched and passed through a colander
4. Add the lobster meat and saute for 2-3 minutes. Add the flour, stirring frequently until light brown.
5. Add tomatoes, ketchup, and coconut milk.
6. Cover the sauce pan and cook until the lobster meat is cooked through (if necessary pour in some water).
7. Stir in whipping cream and remove from the stove.
8. Serve with rice.

L5) Grilled Lobster

Ingredients

1. Lobster	2 kg.
2. Butter melted	1/3 cup
3. Pepper	to taste
4. Salt	to taste.

Method

1. In a large pot boil about 2 gallons of water.
2. Add one live lobster and cover.
3. Cook for 3 minutes. Remove lobster with tongs and return water to boiling.
4. Add 2nd lobster. Cook three minutes and remove with tongs.
5. Turn lobsters upside down and cut lengthwise from the tip of the tail through the head (you may need to use a poultry sheers to cut the tail portion of the shell). This should give you 4 half lobsters.
6. Place lobsters shell side down on a medium hot grill and baste with butter and sprinkle with salt and pepper.
7. Cover grill and cook for about 8-9 minutes or until the meat in the thickest part of the tail turns opaque.
8. Serve with melted butter and a slice of lemon.
9. Serves 2

L6) Boiled Lobster Tails

Ingredients

1. Lobster Tails	4 tails

2.	Bread crumbs finely crushed	2 tbsp
3.	Grated cheese	2 tbsp
4.	Minced fresh parsley	1 ½ tsp
5.	Lemon juice	2 tbsp
6.	White wine	2 tbsp
7.	Garlic crushed	1 clove

Method

1. Make a lengthwise cut through the top of 1 lobster shell, using kitchen shears, and press shell open.
2. Starting at the cut end of the tail, carefully loosen the lobster meat from bottom of the shell, keeping meat attached at end of tail; lift meat through top shell opening, and place on top of shell.
3. Repeat procedure with remaining lobster tails. Place lobster tails on a rack coated with cooking spray, and place rack in a shallow roasting pan.
4. Combine bread crumbs, Parmesan cheese, minced parsley, and paprika in a small bowl; stir in melted butter. Set aside.
5. Combine wine, lemon juice, and garlic; stir well. Brush lobster with half of wine mixture.
6. Broil 5½ inches from heat 9 minutes or until lobster flesh turns opaque, basting with remaining wine mixture after 5 minutes.
7. Sprinkle crumb mixture evenly over lobster, and broil an additional 30 seconds.
8. Serves 4

L7) White Lobster

Ingredients

1. Canned lobster	2 cups
2. Butter	2 tbsp
3. Milk	1 cup
4. Pepper	1 tsp
5. Salt	1 tsp

Method

1. Make a white sauce with the fat, flour and milk.
2. Add lobster cut into small pieces and salt and pepper.
3. Some cooks add a teaspoon of curry powder.
4. Sometimes this preparation is placed in greased or oiled scallop shells.
5. Sprinkled with fine seasoned bread or cracker crumbs and browned in the oven.

L8) Lobster with Oyster in White Sauce

Ingredients

1. Lobster boiled	2
2. Oysters	18
3. Bread crumbs	½ cups
4. White sauce	1 cup
5. Seasoning	to taste.

To make White Sauce:

1. Scald 1 pint of milk. Melt 1, 2 or 3 tablespoons of butter, depending on desired consistency, in a saucepan.

2. Add half a teaspoon of salt, 1, 2, or 4 tablespoons of flour, depending on desired consistency, and let cook for one minute.
3. Stirring constantly. Add the milk little by little, stirring gently until it boils.
4. Let it boil for 3 minutes

Method

1. Cut lobsters in halves.
2. Remove meat and cut in small pieces.
3. Chop oysters. Cook oysters in the white sauce, with the seasonings, for about 2 minutes.
4. Add the lobster meat and the bread crumbs.
5. Fill each lobster shell with the mixture.
6. Put lobster shells in baking pan.
7. Bake in hot oven for about 8 to 10 minutes.
8. Then serve.

L9) Steamed Lobster

Ingredients

1. Lobster cut lengthwise	900 gm
2. Extra virgin oil	¼ cup + 1 tbsp
3. Garlic minced	1 ½ tsp
4. Freshly ground pepper	to taste
5. Dry white wine	¾ cup
6. Salt	to taste.

Method

1. Rub the cut sides of the lobster tails with 1 tablespoon of the olive oil and ½ teaspoon of the garlic.

2. Then season the lobster with salt and pepper.
3. In a large steamer, combine ½ cup of the wine with the lemon juice, juniper berries, bay leaves and thyme sprigs.
4. Add 2 inches
5. Steam the lobsters over high heat about 8 minutes or until the tails are cooked through and the claws are red.
6. Meanwhile, in a large skillet, heat the remaining ¼ cup of oil until shimmering.
7. Add the mushrooms and cook over high heat until tender and just beginning to brown, about 8 minutes.
8. Add the remaining 1 teaspoon of garlic and the chopped thyme and cook until fragrant, about 1 minute.
9. Add the remaining ¼ cup of white wine and cook until almost evaporated.
10. Season with salt and pepper.

L10) Lobster Imperial

Ingredients

1. Marine lobster meat, cooked	1350 gm
2. Bell pepper, finely chopped	1 small
3. Canned pimiento chopped	1 tbsp
4. Dry mustard	1/3 tbsp
5. White pepper	¼ tsp
6. Cayenne pepper	2 dashes
7. Eggs	2

8.	Worcestershire sauce	½ tsp
9.	Paprika	pinch
10.	Salt	½ tbsp

Method

1. Remove any bits of shell from the lobster meat and set aside.
2. Combine the rest of the ingredients and mix.
3. Fold in the lobster meat.
4. Divide among four small ramekins/gratin dishes or baking shells.
5. Lightly spread a coating of mayonnaise on top and sprinkle with paprika.
6. Bake in preheated oven at 350F for 15 minutes or until the mayonnaise browns.
7. Serve hot or cold.

Mussels and Clams

Clams and mussels are very versatile and easy to prepare. *Clams and mussels recipes* are found in different

variations, it can be simply steamed and served as an appetizer or it can be combined in delicious dishes with rice or pasta.

Mussels and clams can be prepared in different ways, including steaming, boiling, baking or frying. Steaming mussels and clams is a wonderful combination. When cooking fresh mussels and clams, pay special attention that they should be alive until you cook them. They are done when their shells open - those that remain closed should be discarded.

M1) Clams Stir-Fry (Teesrya Koshimbiri)

Clams cooked and tossed in a mix of sauted onions, grated coconut and lots of fresh coriander. It is a very light and simple dish. A delicacy in itself.

Cooking time (approx.): 21 minutes

Style: Indian Non-Vegetarian (Gomantak)

Ingredients

1.	Clams washed and cleaned	3 cups
2.	Turmeric powder	½ tsp
3.	Onion Chopped	2 large
4.	Garlic flakes crushed	2
5.	Cumin seeds	1 tsp
6.	Green chillies finely chopped	4
7.	Tomato chopped	1
8.	Grated coconut	1 cup
9.	Garam masala	½ tsp
10.	Coriander leaves chopped	1 cup

11. Oil 2 tbsp
12. Salt to taste

Method

1. In a pot, combine water, clams and salt. Bring to a boil and cook on medium heat for about 15 minutes or till the shells have split open and the clams are tender. Strain and separate the shells retaining the one with the clam and discarding the empty one. Keep aside.
2. Heat the oil in a pan and crackle the cumin seeds. Add the green chillies, garlic and onions. Saute till the onions are transparent. Do not let the onions brown even a little. Add the turmeric powder and mix well. Then add the tomato pieces anf stir-fry briefly.
3. Add the clams, grated coconut and chopped coriander leaves. Mix well and add salt. Stir-fry briefly on high heat for 2 minutes. Sprinkle the garam masala (optional). Cover and cook on low heat for about 4 minutes.

Tips

☆ Fresh clams may need to be repeated and thoroughly washed till all the sand has gone off.

☆ Discard the clams which have not opened after cooking them.

Serve hot: with boiled rice and/or Deep Fried Savory Buns (Vade) accompanied with a tall glass of chilled Appetizing Kokum in Yoghurt (Quick Sol Kadi)

M2) Oyster with Mixed Vegetables

Ingredients

1. Oyster meat	1 kg
2. Onion	500 gm
3. Carrot	200 gm
4. Beetroot	150 gm
5. Green chilli	30 gm
6. Green peas	75 gm
7. Garlic	15 gm
8. Chilli powder	30 gm
9. Turmeric powder	10 gm
10. Pepper powder	10 gm
11. Coriander powder	25 gm
12. Malabar tamarind	25 gm
13. Refined oil	150 gm
14. Curry leaves	few
15. Curry masala powder	20 gm
16. Salt	to taste.

Method

1. Wash the oyster meat and cook in two cups of tamarind extract (item no. 12) with part of pepper and turmeric powder and salt to taste.
2. Grate the ingredients 2 to 6 and semifry in refined oil
3. Grind the spices (7 to 11) and fry it.
4. Fry vegetables and species along with items 14 and 15.

5. Add to the cooked meat and further boil it till gravy thickens.
6. Serve hot

M3) Mussels Vinaigrette

Ingredients

1. Mussels scrubbed and cleaned	48
2. Dry white wine	¾ cup
3. Garlic crushed	3 cloves
4. Mustard	3
5. Salt	1 tsp
6. Sugar	½ tbsp.
7. Dried oregano	½ tbsp
8. Red wine vinegar	6 tbsp
9. Olive oil	1 cup
10. Freshly parsley	few
11. Onion chopped	1

Method

1. Steam mussels in white wine and 2 cloves garlic until they open.
2. Drain and cool.
3. In a large bowl or big jar, mix together the French mustard, remaining garlic, salt and pepper, sugar, oregano and vinegar.
4. Add olive oil and mix again.
5. In a serving dish, toss mussels with ¾ cup of this dressing.
6. Garnish with parsley, dill, onion.

7. Chill and serve.
8. Serves 6

M4) Sauced Mussel

Ingredients

1. Mussels
2. White Wine Vinegar
3. Dry White wine or Sherry

Method

1. Steam open mussels by the method known as "Moules Mariniere"
2. This involves quickly steaming open the shells in a wide based covered pan with approx. 2.5cm water.
3. Reserve the cooking liquid and remove mussels from the shells.
4. Pack the mussels into a screw-top jar and add reserved cooking liquid to half fill the jar.
5. Add white wine vinegar to another quarter of the jar and the final quarter with a dry white wine or sherry.
6. Seal the jar and place in the refrigerator.

M5) Mussel Casserole

Ingredients

1. Mussels	48
2. Butter	1 tbsp
3. Cream of mushroom	250 gm
4. Sherry	2 tbsp
5. Black pepper freshly ground	2 tbsp

Method

1. Remove meat from shells by steaming.
2. Grease a casserole dish with the butter and place in mussel meat.
3. Top with soup and add the sherry.
4. Sprinkle with freshly ground black pepper.
5. Bake in a moderate oven for 25 minutes.
6. Serve immediately with rice or vegetables.
7. Serves 4

M6) Easy Mussel

Ingredients

1.	Fresh mussels	900 gm
2.	Butter	1 tbsp
3.	Lemon juice	to taste
4.	Salt	to taste
5.	Garlic minced	2 cloves

Method

1. Discard mussels that are already opened.
2. Put a little water in a big pot and bring to a boil. Dump in the mussels, salt and cook until mussels open.
3. Remove mussels from heat and start pulling them from their shells.
4. In a bowl melt the butter and garlic in the microwave for 45 secs. Pour over mussels and squeeze lemon juice.

M7) Steamed Mussels with Creamy Fish Filling

Ingredients

1.	Large mussels	24
2.	Shrimp paste	1 tsp
3.	Whitefish fillets chopped	250 gm
4.	Egg white	1
5.	Onion finely chopped	1 small
6.	Fresh ginger finely chopped	2 tsp
7.	Garlic crushed	1 clove
8.	Cream	1 tbsp
9.	Fresh basil chopped	1 tbsp
10.	Red chillies sliced	2
11.	Fresh lemon grass finely chopped	1 stem

Method

1. Scrub mussels and remove beards. Place mussels in pan, cover with cold water, and bring to boil.
2. Simmer, covered, for about 3 minutes until mussels begin to open.
3. Drain mussels, rinse under cold water; drain well. Remove and discard top shell from each mussel.
4. Heat oil in pan; add onion, ginger, garlic and lemon grass, cook, stirring, until onion is soft.
5. Add paste, cook, and stirring, for 1 minute; cool.
6. Blend or process fish, egg white and cream until smooth.
7. Combine fish mixture, onion mixture and basil in bowl; mix well.

8. Spoon mixture onto mussel in shells, smooth surface, top with chilli.
9. Just before serving, place mussels in bamboo steamer in single layer.
10. Cook, covered tightly, over pan of boiling water, for about 3 minutes or until fish mixture is cooked through.
11. Serves 6

M8) Frilled Mussel with Hot Fish Sauce

Ingredients

1.	Fresh mussels	900 gm
2.	Fish sauce	3 tbsp
3.	Lemon juice	1 lemon
4.	Sugar	1 tsp
5.	Red pepper	2
6.	Garlic	2 cloves
7.	Water	2 tbsp

Method

1. Clean and scrub the mussels.
2. Prepare sauce: Blend garlic and red peppers into a paste.
3. Add lemon juice, fish sauce, sugar, and water.
4. Preheat broiler.
5. Place clean, dry mussels in a single layer on a baking sheet.
6. Broil mussels just until they are all open and aromatic.

7. Remove from broiler and serve with prepared sauce and rice.
8. Serves 4

Crabs

Crab is another delicious seafood option and you can find innumerous *crab recipes*. Whole crabs and crabmeat can be prepared in a variety of ways including frying, steaming, broiling or in soups. To cook crabs simply throw them in a pot of boiling water and cook them for twenty minutes for crabs up to 2 lb and add five minutes per extra pound. The shell of some varieties turns red when they are done.

Cr1) Spicy Baby Crabs

Baby crabs simmered with spices and tomatoes. A semi-dry preparation with the gravy almost clinging to the baby crabs making them succulent and lip-smacking.

Style: South Indian Non-Vegetarian

Cooking time (approx.): 13 minutes

Ingredients

1.	Baby crabs washed and cleaned	8
2.	Mustard seeds	1 tsp
3.	Green Chillies finely chopped	4
4.	Onion Chopped	1 large
5.	Ginger–garlic paste	2 tsp
6.	Red chilli powder	1 tsp
7.	Turmeric powder	½ tsp
8.	Tamarind paste	1 tsp
9.	Oil	2 tbsp
10.	Salt	to taste
11.	Coriander leaves chopped	to garnish

Method

1. Wash the crabs thoroughly and cut each into two pieces.
2. Heat the oil in a pan and crackle the mustard seeds. Add the green chillies, curry leaves, onions and saute till the onions are transparent. Add the ginger-garlic paste and saute for about 3 minutes or till the onions are light brown.
3. Add the chopped tomatoes, red chilli powder and turmeric powder. Stir well and fry till the oil separates. Add the baby crabs, tamarind paste and salt. Add a little water if required and mix well. Cover and cook on low heat for about 10 minutes. Garnish with finely chopped coriander leaves.

Tips

- ☆ Big crabs can be used in this recipe. However, they need to be cooked in some water first. Also, their big claws have to be detached and cracked before cooking.
- ☆ Any souring agent can be used in place of tamarind paste.

Serves: 4. Serve hot with: plain boiled rice and/or warm.

Cr2) Baked Crabs

Ingredients

1.	Crabs medium	4
2.	Onion minced	1
3.	Curry powder	1 desert spoon
4.	Green chillies Chopped	1–2
5.	Garlic chopped	3–4 pods
6.	Ginger chopped	¼" piece
7.	Bread slice	1
8.	Salt	to taste

Method

1. Plunge the crabs in boiling salted water.
2. Remove the flesh from the shell and set aside.
3. Brown the onion in oil.
4. Add the curry powder, green chillies, ginger, garlic and bread.
5. Add the flesh of the crabs.
6. Cook till dry.

7. Fill back into crab shells and brown under hot grill
8. Serve hot.

Cr3) Crab Stuffed Mushrooms

Ingredients

1.	Crab meat	225 gm
2.	Fresh mushrooms	35–40
3.	Celery finely chopped	¼ cup
4.	Onion finely chopped	2 tbsp
5.	Red bell pepper finely chopped	3 tbsp
6.	Green pepper	¼ cup
7.	Oyster cracks crushed	2 cups
8.	Garlic crushed	1 tbsp
9.	Old Bay seasoning	1 tsp
10.	Black pepper ground	¼ tsp
11.	Egg	1
12.	White cheddar cheese slices	6
13.	Salt	¼ tsp

Method

1. Preheat oven to 400 degrees.
2. Sauté celery, onion, and pepper in butter for 2 minutes, transfer to a plate and cool in the refrigerator.
3. While vegetables cool, wash mushrooms and remove stems.
4. Set caps to the side and finely chop half of the stems.
5. Discard the other half of the stems or use elsewhere.

6. Combine the sautéd vegetables, chopped mushroom stems, and all other ingredients (except cheese slices) and mix well.
7. Place mushroom caps in individual buttered casseroles or baking dishes (about 6 caps each).
8. Spoon 1 teaspoon of stuffing into each mushroom.
9. Cover with sliced cheese.
10. Bake in oven for 12-15 minutes until cheese is lightly browned.
11. Serves 6

Cr4) Chinese King Crab with Mushroom and Snow Peas

Ingredients

1.	King crab meat	500 gm
2.	Mushroom, thinly sliced	300 gm
3.	Snow peas	100 gm
4.	White pepper	½ tsp
5.	Pea nut oil	3 tbsp
6.	Corn flour	1 tbsp
7.	Garlic crushed	1 clove
8.	Ginger ground	½ tsp
9.	Chicken stock	1 cup
10.	Shallots, finely chopped	2

Method

1. Steam snow pea for 5 minutes.
2. Place in bottom of serving dish and keep warm.
3. In a large fry pan, heat oil until very hot but not smoking.

4. Add garlic a sauté until golden.
5. Remove. Add crab meat and toss, cooking for 1 minute.
6. Stir in mushrooms and cook for 1 minute more.
7. Add Stock and pepper and bring to a boil.
8. Corn flour, dissolved in ½ cup Cold water
9. Add cornstarch mixture and cook, stirring until sauce thickens.
10. Pour over snow pea pods, sprinkle with ginger, and toss to mix.
11. Sprinkle with shallots and serve hot with cooked rice.

Cr5) Crab Monterey

Ingredients

1.	Crab meat, drained and flaked	250 gm
2.	Frozen chopped spinach	600 gm
3.	Chicken stock powder	2 tsp
4.	Sour cream	250 gm
5.	Cayenne pepper	2 tsp
6.	Artichoke hearts, drained and quartered	250 gm
7.	Butter	1 tsp

Method

1. Place frozen spinach in a 10-cup shallow microwave-safe casserole;
2. Sprinkle with instant chicken broth.
3. Cover with wax paper.

4. Defrost in microwave oven ten minutes, rotating dish and stirring once.
5. Drain well; stir in sour cream and cayenne.
6. Arrange artichoke hearts and crab atop.
7. Dot with butter and cover again with wax paper.
8. Microwave on high 5 minutes, or until heated through.

Cr6) Crab Meat Supreme

Ingredients

1.	Crab meat	1 Kg
2.	Onion chopped	1 cup
3.	Green capsicum chopped	1 cup
4.	Celery chopped	1 cup
5.	Butter	500 gm
6.	Parsley chopped	½ cup
7.	Tabasco sauce	1 tsp
8.	Lemon juice	2 lemon
9.	Pimento canned drained	60 gm
10.	Canned Mushroom small drained	1500 gm
11.	Sherry	1 cup
12.	Flour	1 cup
13.	Pepper	1 tsp
14.	Milk	4 cup
15.	Savory biscuit crumbs	1 cup
16.	Salt	4 tsp

Method

1. Cook onion, pepper and celery in half of the butter; do not brown.
2. Remove from the heat and combine with crab, parsley, 2 teaspoons salt.
3. Tabasco, lemon juice, pimentos, mushrooms, and sherry. Combine flour, remaining salt, and pepper in double boiler.
4. Slowly stir in milk and cook until mixture is smooth, stirring.
5. Remove from heat and stir in butter until melted.
6. Combine with crab mixture. Pour into casserole and sprinkle with crumbs.
7. Bake in a preheated 325F. Oven for 20 minutes.
8. Serves 16

Squid

S1) Baked Stuffed Squid

Ingredients

1.	Squid	450 gm
2.	Olive oil	6 tbsp
3.	Fresh parsley	½ cup
4.	Fresh mint leaves	¼ cup
5.	Black raisins	¼ cup
6.	Pepper ground	½ tsp
7.	Peeled Tomato, drained	4
8.	White wine	1/3 cup
9.	Salt	to taste
10.	Basmati rice	1/3 cup

Method

1. Wash and clean the squid, separating the outer sacs from the heads and tentacles, removing and discarding the translucent cartilage, and small sand bag and ink.
2. Rub salt on the outer sacs and rinse them inside and out with cold water.
3. Heads and tentacles should be rinsed thoroughly and cooked along with the sacs after you stuff the latter.
4. Drain and set aside.
5. Heat ¼ cup of the oil in a heavy frying pan, then add the onion and cook, without browning, until transparent.
6. Stir in the rice and saute a few minutes, until golden.
7. Blend in the parsley, mint, 2 tablespoons wine, pine nuts, and raisins, and season with salt and pepper to taste.
8. Add enough water to half cover and cook for a few minutes, then stuff the squid sacs with the mixture using a very small spoon and allowing enough liquid in each for the rice to cook.
9. Seal opening with skewers or toothpicks.
10. Place the stuffed sacs with the heads and tentacles in a baking-serving dish.
11. Sprinkle with salt and pepper and set aside.
12. Meanwhile, combine the tomatoes, 1/3 cup wine, and a little salt and pepper in a small saucepan and simmer for 5 minutes.

13. Pour the sauce over the squid and dribble the remaining 2 tablespoons olive oil over the top.
14. Bake in a medium-slow oven (300 F) for 1 ½ hours or until the squid and rice are tender and the sauce has thickened.
15. Serve warm or cold.

S2) Batter Fried Squid

Ingredients

1.	Squid	6
2.	Eggs	2
3.	Flour	½ cup
4.	Olive oil	¾ cup
5.	Salt	pinch

Method

1. Wash cleaned squid; remove tentacles; dry fish thoroughly.
2. Cut the body of squid into rings, dip them in flour, then in egg beaten with a pinch of salt.
3. Fry rings in very hot oil. The fish is done as soon as it is browned on all sides.
4. Frying time should in no case exceed 5 minutes because squid toughens when overcooked.

S3) Squid with Garlic and Tomato

Ingredients

1.	Small fresh squid	500 gm
2.	Vinegar	½ cup
3.	Onion sliced	1 medium

4. Tomato chopped 2 medium
5. Pepper to taste
6. Monosodium glutamate (vet-in) 1 tsp
7. Salt to taste.
8. Garlic 10 cloves

Method

1. Wash the squids very well. Remove the long thin membrane in the head and slit the eyes to bring out the ink.
2. Place the squids in a saucepan with vinegar, 6 cloves garlic crushed, salt and pepper.
3. Cover and cook slowly until the squids are tender.
4. Cut cooked squids into ½ inch slices crosswise.
5. Crush remaining garlic and saute in a little lard in another pan.
6. Add the onion and tomatoes and cook until tomatoes are very soft.
7. Add the squids and the liquid in which they were boiled.
8. Simmer for 7 minutes.
9. Season with salt, pepper and vet-in.

Chapter 4
Dry Fish Recipes

D1) Dried Anchoviella Fried

Ingredients

1.	Dried Anchoviella	200 gm
2.	Onion (Small)	50 gm
3.	Turmeric powder	¼ tsp
4.	Red Chillies	10
5.	Ginger	a tiny piece
6.	Oil	to fry

Method

1. Keep the fish in cold water for 10 minutes and remove the head and tail.
2. Grind nicely the turmeric, chillies, ginger and onion.
3. Paste the masala on the clean ed fish.

4. Add salt, if needed
5. Keep it for 10- 20 minutes.
6. Fry it in oil for 5–10 minutes till the fish become crisp.

D2) Dried Anchoviella Thorn

Ingredients

1.	Dried fish	200 gms
2.	Coconut	¼ (scraped)
3.	Green chillies	5 Nos.
4.	Onion	25 gms
5.	Ginger	1 small piece
6.	Pepper	¼ tsp
7.	Mustard	1 tsp
8.	Coconut oil	
9.	Black gram Dal	1 tsp

Method

1. Keep the fish in cold water for 10 minutes and remove the head and tail.
2. Cook fish in little water and remove the bones.
3. Smash into smaller pieces.
4. Fry mustard seed and dal till the mustard burst.
5. Then add sliced onion, ginger and green chillies.
6. Fry till it is brown.
7. Then add smashed fish and scrape coconut and sprinkle the powdered pepper.
8. Add salt, if necessary.
9. Fry few minutes till the water evaporates.

D3) Dried Anchoviella Chutney

Ingredients

1.	Dried Anchoviella fish	200 gms
2.	Onion	2 Nos
3.	Red chillies	10 Nos
4.	Pepper grains	
5.	Ginger	a tiny piece
6.	Coconut	½ No
7.	Coconut oil	2 tsp.

Method

1. Keep the dried fish in water for 10–20 minutes.
2. Remove the head, tail and bones
3. Fry the flesh portion in little coconut oil till the water contents evaporates.
4. Fry the scraped and ginger, onion red chillies for 5 minutes
5. Grind it with Pepper and ginger.
6. Mince the fried fish and mix well with the ground masala.
7. Add salt if needed
8. Pour little coconut oil
9. Tamarind can also be added for good look.

D4) Dried Fish Pachadi

Ingredients

1.	Dried fish and powdered	½ Cup
2.	Small onion sliced	¼ cup
3.	Ginger sliced	1 ½ tsp

4. Red chilli ground ½ cup
5. Mustard ¼ tsp
6. Finely chopped onion 1 tsp
7. Curry leaves few
8. Curd 1 cup

Method

1. Fry ingredients 2–4 in oil.
2. Mix with dry fish powder
3. Add 1 cup of curd.
4. Fry ingredients 5–7 in oil .
5. Add this to the curd prepared above.

D5) Dried Shark Fried

Ingredients

1. Dried shark 250 gms
2. Onion (Small) 50 gms
3. Turmeric powder ¼ tsp
4. Red chillies 10 Nos
5. Ginger 1 tiny piece
6. Oil to fry.

Method

1. Keep the dried shark in water for 20–30 minutes to desalt the fish.
2. Remove the skin and bone and cut the flesh into pieces of 2 cm × 1 cm approximately or to convenient size.
3. Grind nicely the turmeric, chillies, ginger and onion.

4. Smear the masala on the cleaned fish pieces.
5. Add salt if necessary.
6. Fry it in oil for 5 to 10 minutes till the fish becomes brown.

D6) Dried Silver Belly

Ingredients

1. Fish (Silver belly)	250 gm
2. Red chillies	10 nos.
3. Coconut oil	10 tsp
4. Pepper corns	2–3
5. Curry leaves	few

Method

1. Soak the dried silver bellies in water for 6 minutes.
2. Remove head and tail.
3. Fry it thoroughly in Coconut oil.
4. Fry dried curry leaves, red chillies and 2–3 pepper corns.
5. Mix both the fried ingredients and powder it.
6. Store in polythene begs.
7. Mix it with coconut oil and Serve.

D7) Dried Saurida–Cooked and Fried

Ingredients

1. Fish	200 gm
2. Green chilli	4 Nos.
3. Small onion.	10 nos.
4. Red chilli	8 Nos.

5. Turmeric — a bit
6. Salt — to taste

Method

1. Soak the fish in water for 10 minutes.
2. Remove skin and tail portion.
3. Cut into two pieces.
4. Grind all the ingredients together to a paste and smear it on the fish.
5. Semi cook in little water.
6. Once the water dries up add coconut oil and fry it.

D8) Dried Ray-fried

Ingredients

1. Cleaned fish — 250 gm
2. Red chillies — 12 Nos.
3. Small onion. — 8 Nos.
4. Scraped coconut — ½

Method

1. Soak the ray pieces in water to remove excess salt.
2. Then smash it into almost fiber–like pieces.
3. Fry ingredients 2 to 4 slightly powder it
4. Mix with ray fibers and sauté in hot coconut oil till the fish fibers gets fried.

D9) Fried and Dried Bombay Duck

Ingredients

1. Bombay duck (Dried) — 8

2.	Turmeric powder	1 tsp
3.	Chilli powder	¼ tsp
4.	Salt	to taste.
5.	Cumin	¼ tsp
6.	Garlic	2 flakes
7.	Peppercorns	3
8.	Tamarind	whole nut size ball
9.	Ginger	¼ " piece

Method

1. Soak the Bombay duck in water.
2. Remove the central bone.
3. Squeeze out all the water.
4. Grind together the chillies, cumin, garlic, pepper corn, tamarind, and ginger in vinegar and salt.
5. Stuff Bombay duck with above masala
6. Secure the fish with string and fry till both sides are brown.
7. Remove the string and serve.
8. Serve 4

Chapter 5
Pickles

PI1) Fish Pickle

Ingredients

1.	Fish meat	1 kg
2.	Oil	500 gms
3.	Green chillies	100 gms
4.	Ginger	50 gms
5.	Garlic	200 gms
6.	Curry leaves	50 gms
7.	Mustard seeds	50 gms
8.	Chilli powder	100 gms
9.	Turmeric powder	100 gms
10.	Salt	150 gms
11.	Pepper powder	50 gms

12.	Vinegar	600 ml
13.	Fenugreek	25 gms

Method

1. Mix the fish meat with pepper powder, salt and turmeric powder and keep aside for ½ an hour.
2. Half part of chilli powder, garlic and ginger is ground and keep aside.
3. Fry minced meat in oil.
4. In the remaining oil, semi-fry the remaining part of chilli powder, garlic , ginger and mustard seeds, green chillies and curry leaves together with the already ground masala.
5. Add vinegar and fenugreek powder.
6. Mix the fried fish kheema and allow the contents to boil for 15 minutes.
7. Let it cool before bottling.

PI2) Oyster Pickle

Ingredients

1.	Oyster meat	1 kg
2.	Groundnut oil	250 gm
3.	Green Chilli	100 gm
4.	Ginger	50 gm
5.	Garlic	20 gm
6.	Chilli powder	150 gm
7.	Turmeric powder	30 gm
8.	Fenugreek powder	20 gm
9.	Pepper powder	20 gm

10.	Salt	30 gm
11.	Curry leaves	few
12.	Mustard seed (as flakes)	50 gm
13.	Vinegar	700 ml

Method

1. Wash the oyster meat and mix it with a portion of pepper and turmeric powder and salt to taste.
2. Keep this for one hour and then fry in groundnut oil.
3. Grind chilli powder and half of the green chilli, garlic and ginger.
4. Grate the second half of the green chilli, garlic and ginger and add other spices and semi-fry in oil with curry leaves and mustard seed.
5. Mix it with the ground masala and fry till the colour turns to brown.
6. Fry the oyster meat with the above mixture and boil it with vinegar.

PI3) Mussel Pickle

Ingredients

1.	Muscle meat	1 kg
2.	Ground nut oil	250 gm
3.	Green chilli	100 gm
4.	Ginger	50 gm
5.	Garlic	20 gm
6.	Chilli powder	150 gm
7.	Turmeric powder	30 gm
8.	Fenugreek powder	20 gm

9.	Pepper powder	20 gm
10.	Salt	30 gm
11.	Curry leaves	few
12.	Mustard seed	50 gm
13.	Vinegar	700 ml

Method

1. Wash the mussel meat and mix it with a portion of pepper and turmeric powder and salt to taste.
2. Keep this mix for one hour and then fry in groundnut oil.
3. Grind chilli powder and half part of the green chilli, Garlic and Ginger.
4. Grate the second half of the green chilli, garlic and ginger and add other spices (6 to 9) and semi fry in oil with curry leaves and mustard seed.
5. Mix it with the ground masala and fry till the colour turns to brown.
6. Fry the mussel meat with the above mixture and boil it with vinegar.
7. Bottle the product after 3-5 days of curing.

PI4) Salted Fish Pickle–Parra

Ingredients

1.	Salted fish or Mackeral	50 pieces
2.	Cumin	1 ½ cup
3.	Pepper corns	½ cup
4.	Garlic	4 whole pods
5.	Ginger	6" piece

6.	Chilli powder	4 cups
7.	Turmeric	1 ½ cups
8.	Vinegar	300 ml

Method

1. Wash salted fish pieces in vinegar sprinkled with turmeric powder.
2. Dry in Sun for a day.
3. Grind the cumin, pepper corns 2 pods of garlic and approximately half of the ginger in vinegar.
4. Mix the chilli powder and turmeric.
5. Crush the remaining 2 pods of garlic and ginger.
6. Mix this with the ground masala and the fish.
7. Put it in earthenware and cover the fish with vinegar.
8. Pickle for about a month, at least, before using.
9. Before serving, take out the mackerel of fish and shallow fry quickly on both sides for a minute or two.

PI5) Dry Bombay Duck Pickle

Ingredients

1.	Bombay ducks large	12
2.	Garlic	1 pod
3.	Cumin	1 tsp
4.	Ginger	2" piece
5.	Turmeric powder	1 tbsp.
6.	Cinnamon	1" piece
7.	Cloves	3

8.	Vinegar	½ cup
9.	Oil	½ cup
10.	Salt	to taste.

Method

1. Clean the Bombay ducks and cut into pieces.
2. Fry and set aside.
3. Grind the garlic, ginger, cumin, turmeric, cinnamon, and cloves in vinegar.
4. Fry till brown in oil.
5. Add the Bombay dicks, cook, cool briefly and bottle.

Chapter 6
Curry

Curries made of fresh fish are really delicious and are made in Kerala homes along the coastal areas almost everyday. Many people cannot have their daily meal with out a fish or prawn dish.

CU1) Dried Anchoviella Curry

Ingredients

1. Dried Anchoviella curry	200 gms
2. Onion (Small)	25 gms
3. Red Chillies	10 Nos.
4. Green chillies	2 Nos
5. Coconut	¼ No
6. Mustard seed	½ tsp
7. Curry leaves	few
8. Turmeric	½ tsp

9.	Coriander	½ tsp
10.	Ginger	a tiny piece

Method

1. Soak the dried Anchoviella in water for 5–10 minutes.
2. Remove the head and tail of the fish and wash it thoroughly.
3. Grind nicely chilli, turmeric, coriander, ginger.
4. Fry the mustard till it burst.
5. Add sliced onion green chillies and fry.
6. Add masala and warm it for 3 minutes.
7. Pour sufficient water and boil it.
8. Add cleaned fish
9. Boil it for 5–10 minutes till it is cooked.
10. Add salt to taste
11. Add coconut milk and boil for 2 minutes.

CU2) Dried Shark Curry

Ingredients

1.	Dried shark	250 gms
2.	Coriander powder	½ tbsp
3.	Coconut	½
4.	Green chillies	10
5.	Red chillies	10
6.	Ginger	1 small piece
7.	Onion small	50 gms
8.	Tamarind	little

Method

1. Soak the dried shark in water for 20–30 minutes to desalt the fish.
2. Remove the skin and bones.
3. Cut the flesh into pieces of 2 cm x 1 cm approximately or to any convenient size.
4. Fry coriander powder, red chillies and coconut in a little quantity of oil till they become brown and grind well.
5. Mix the cleaned fish with the ground masala.
6. Add tamarind juice and boil in sufficient quantity of water for 20–25 minutes.
7. Add sliced green chillies, onion and ginger.
8. When the gravy becomes thick, add some onion pieces fried in oil and remove.

CU3) Dried Saurida Curry

Ingredients

1. Fish	400 gm
2. Coriander	a bit
3. Turmeric	1 tsp
4. Red chillies	10
5. Coconut	½
6. Ginger	one piece
7. Green Chillies	5 nos.
8. Curry leaves	few
9. Tamind piece	3

Method

1. Soak the fish 10–3-minutes in water.
2. Skin it and remove tail portion and cut into pieces.
3. Grind coriander to a paste.
4. Chop Green chillies, Ginger.
5. Mix fish, masala and chopped ingredients and add sufficient water and cook.
6. When it becomes almost semi-solid pour coconut oil over it.

CU4) Fish Curry in Coconut Milk

Ingredients

1.	Fish	500 gms
2.	Chillies	100 gms
3.	Pepper corn	100 gms
4.	Corriander seeds	100 gms
5.	Sweet cumin	100 gms
6.	Cumin	100 gms
7.	Fenugreek	2 tsp
8.	Cinnamon	1″ thick
9.	Rampe	1″
10.	Salt	2 tsp
11.	Cardamom	2 pieces crushed
12.	Coconut milk	1 cup
13.	Goraka	1 tsp
14.	Garlic	4 cloves
15.	Ginger	1 slice
16.	Onion	1 large chopped.

Method

1. Grind ingredients (2–6) into a fine powder.
2. Make of ingredients no 13–16 to a fine paste.
3. Mix the above powdered and paste and ingredients (7–11) along with coconut milk.
4. Wash the fish and add to the gravy mixture.
5. Allow the entire mixture to boil and simmer on low flame until gravy thickens.

CU5) Cuttle Fish Curry

Ingredients

1.	Cuttle Fish (Cleaned)	500 gms
2.	Coriander	1 tsp
3.	Chilli powder	2 tsp
4.	Onion (Sliced)	100 gms
5.	Mustard	½ tsp
6.	Tomato	200 gms
7.	Green chillies	5 nos
8.	Ginger	1″ piece
9.	Red chilli	3 nos
10.	Salt	to taste

Method

1. Slice cuttle fish into small pieces and boil thoroughly.
2. Grind coriander, chilli powder, onion, green chillies ginger etc.
3. Fry mustard and red chilli till the mustard burst.
4. Add slice tomato and fry for 3 minutes.

5. Add ground masala with cooked cuttle fish and sufficient water.
6. Add salt to taste and allow to boil.

CU6) Coorgi Fish Curry

Ingredients

1.	Medium sized fish	one
2.	Kauli channa (White grams)	120 gms
3.	Blanched, finely sliced almonds	60 gms
4.	Sliced raisings	60 gms
5.	Green chillies	4
6.	Oil	2 tbsp
7.	Ginger	1″ piece
8.	Garlic	4 pods
9.	Salt	to taste
10.	Chilli powder	to taste
11.	Coriander leaves	to garnish.

Method

1. Wash and soak the grams in water overnight.
2. Next morning, boil them in the water they were soaked in until dry and soft.
3. Mash coarsely.
4. Grind ginger, garlic and coriander leaves to a paste.
5. Fry all the above ingredients (except fish) until they absorb all the oil.
6. Clean, scale and gut the fish. Then slit on one side and stuff the mixture into the fish.

7. Tie with a thread and lightly brush butter all over.
8. Place the fish on the invertible grill rack (facing upward) and put the rack inside the grill chamber.
9. Shut the glass door and turn the grill knob to sim position.
10. Keep turning the fish till both sides are evenly browned. Remove the thread before serving.

CU7) Fish Masala Curry

Ingredients

1.	Fish dressed as for frying	500 gm
2.	Fish Masala powder	1 desert spoon
3.	Sliced onion	2 desert spoon
4.	Green Chillies split	2 nos.
5.	Small variety garlic	9 pods
6.	Ginger chopped lengthwise	2 tsp
7.	Medium size ripe tomato	1 (Cit into 4 pieces)
8.	10 gm of tamarind extract in water	1 cup
9.	Salt	to taste

Method

Mix all the above ingredients together and boil till the gravy thickens.

Curry is ready to serve

CU8) Caldine (Pomfret Curry)

Ingredients

1.	Pomfret (medium)	1

2.	Coconut (Scrapped)	½
3.	Cumin	¼ tsp
4.	Turmeric powder	1 tsp
5.	Garlic	2–3 flakes
6.	Onion medium	(Chopped) 1
7.	Ginger	½ " piece
8.	Green chillies	2–3
9.	Tamarind	to taste
10.	Salt	to taste
11.	Coriander leaves	for garnishing

Method

1. Soak the coconut in boiling water. When cool, pass through blender.
2. Strain the milk and set aside.
3. Add more warm water, cumin, turmeric, garlic, and ginger to coconut.
4. Pass through blender once again.
5. Strain and set thin juice aside.
6. Cut fish into slice and fry if desired.
7. Add the thick and thin juices of the coconut.
8. Cut the green chillies length-wise.
9. When this comes to a boil, add the fish and cook till done.
10. Finally add tamarind and salt to taste
11. Garnish with coriander.

CU9) Prawn Curry with Bhindi

Ingredients

1.	Prawns (Shelled and deveined)	500 gm
2.	Fresh coconut (Scrapped)	½
3.	Kashmiri chillies	6
4.	Cumin	½ tsp
5.	Pepper corns	¼ tsp
6.	Whole coriander	½ tbsp
7.	Turmeric	1 ½ tsp
8.	Ginger (Chopped)	½ " piece
9.	Garlic	4 flakes
10.	Tamarind	Small ball
11.	Onion	1 large
12.	Salt	to taste.
13.	Bhindis	12

Method

1. Grind fine the coconut, chillies, cumin, pepper, coriander, ginger, garlic and tamarind.
2. Add turmeric.
3. Fry the onion till brown.
4. Add the ground masala and fry well.
5. Add prawns at this stage.
6. Cover with water and cook till done.
7. Add bhindies along with the prawns on the end.

CU10) Amotik–(Cuttlefish-Sour-Hot Curry)

Ingredients

1. Cuttle fish (cleaned)	500 gm
2. Kashmiri chillies	8
3. Cumin	½ tsp
4. Pepper corn	½ tsp
5. Ginger	½" piece
6. Garlic	2–3 pods
7. Tamarind	2 small ball
8 Onion (sliced)	1 small
9. Vinegar	to taste
10. Salt	to taste

Method

1. Grind the chillies, pepper ginger, garlic and tamarind together in vinegar.
2. Fry an onion.
3. Add the ground masala fish and salt.
4. Cover with water and cook till the cuttle fish is done.

CU11) Tomato–Prawn Curry

Ingredients

1. Prawns (Cleaned and deveined)	1 cup
2. Dry Kashmiri chillies	20
3. Coriander	1 tbsp
4. Cumin	1 tsp
5. Big Tomato (whole)	25

6.	Coconut (Scrapped)	½
7.	Onion (Sliced)	12
8.	Garlic (Chopped)	6 pods
9.	Ginger (Chopped)	1" piece
10.	Green chillies	5
11.	Curry leaves	few
12	Coriander leaves (fresh)	few
13.	Pepper	to taste
14.	Salt	to taste.

Method

1. Grind the red chillies, coriander and pepper to a paste and set aside.
2. Steep the tomatoes in hot water and cook for a while.
3. Strain the juice and set aside.
4. Extract juice from the coconut and set aside also.
5. Put some Ghee in a pan and fry the sliced onions garlic and ginger.
6. Add the prawns and salt to this and fry well.
7. Add the ground masala, green chillies curry and coriander leaves, coconut milk and tomato juice.
8. Cook on a slow fire till it thickness.

CU12) Prawn Malai Curry

Ingredients

1.	Prawns	500 gm
2.	Ginger	1" piece
3.	Garlic	3 pods.

4.	Coconut milk	½ cup
5.	Curds	40 gm
6.	Kaju	5 Nos.
7.	Red Chillies powder	½ tsp
8.	Turmeric powder	½ TSP
9.	Tomato	3
10.	Tomato paste	3 tsp
11.	Onion	1 small
12.	Garam masala	to taste
13.	Jayfar	pinch
14.	Javantri	pinch
15.	Jeera powder	5 gm
16	Mustard oil	

Method

1. Take mustard oil in a small bowl and heat.
2. Stir all the ingredients and paste in this bowl on low flame.
3. Take boiled prawn and add it to 2–3 tsp tomato paste, as much as to make it thick and stir well.
4. At the end of this put garam masla, Javantri, Jayfal and Jeera powder in it.
5. Serve hot.

CU13) Prawn Curry (South Indian Cuisine)

Ingredients

1.	Prawn	500 gms (before cleaning)
2.	Oil	2 tsp

3.	Cumin Seeds	½ tsp
4.	Ginger Garlic paste	½ tsp
5.	Clove	1
6.	Cinnamon	1 "
7.	Fennel Seeds	½ tsp
8.	Onions	2 medium
9.	Tomatoes	2 medium
10.	Salt	to taste
11.	Turmeric	a pinch
12.	Chilli powder	1 tsp
13.	Coriander powder	1 tsp
14.	Water	250 ml
15.	Coconut Milk	100 ml

Method

1. Clean the prawns and let it drain.
2. Heat a pan with oil,
3. Add cumin seeds, then add the ginger garlic paste to the oil.
4. Fry for 1 min, then add cloves, cinnamon and fennel seeds.
5. Fry for 1 min
6. Then add finely chopped onions.
7. Sauté till they turn pink.
8. Then add tomatoes.
9. Fry till they become mushy.
10. Then add the prawns, Fry on high flame till they are well mixed.

11. Add salt and turmeric. Simmer for 5 mins. Water will come out of the prawns,
12. Fry till it evaporates.
13. Then add the chilli powder, Coriander powder. Simmer for 5 mins.
14. Oil should come out by now.
15. Then add 250 ml water. Bring to boil. Let it boil for 10 mins, by when the water will go down to 100 ml. It depends on the thickness you want in the gravy.
16. Then finally add the thick coconut milk and bring to boil.
17. Garnish with coriander leaves. Serve with pooris.

CU14) Prawn Curry with Coconut

Ingredients

1.	Prawns (peeled)	450 gm
2.	Frozen peas	100 gm
3.	Coconut (desiccated)	50 gm
4.	Tamarind	25 pods
5.	Milk of coconut	400 ml
6.	Curry leaves	4–5
7.	Green chilli Seeded and cut half	1
8.	Ginger	40 gm (1 x 5piece)
9.	Ginger paste	1 tsp
10.	Ghee	25 gm
11.	Tomato (skinned and diced)	2

12.	Fenugreek seeds	1 tsp
13.	Coriander seeds	2 tsp
14.	Red chillies	2 small
15.	Onion large and chopped	1
16.	Salt	to taste.

Method

1. Heat oil in a pan and season with mustard seeds.
2. Stir over medium heat until they begin to splutter.
3. Add garlic and ginger and stir for a minute.
4. Add green chillies and stir lightly.
5. Add the onions and sauté until translucent and glossy.
6. Add turmeric powder and stir well.
7. Add prawn and the third extract of coconut milk. Bring it to a boil. Lower the heat and simmer, turning carefully once, for three minutes.
8. Add salt and stir well. Add curry leaves, tomatoes and the second extract.
9. Cover and simmer for three minutes.
10. Remove the pan from heat and gently (to make sure that the fillets do not break) stir-in the first extract of coconut milk.
11. Return the pan to heat and bring to a boil over low heat.
12. Sprinkle lemon juice and stir carefully. Remove from heat and adjust the seasoning.
13. Remove to a bowl and serve with steamed rice.

CU15) Burmese Coconut Prawn Curry

Ingredients

1.	Prawns	750 gm
2.	Turmeric ground	1 tsp
3.	Onion Chopped	1 cup
4.	Garlic crushed	4 cloves
5.	Paprika	½ tsp
6.	Red Chillies powder	1 tsp
7.	Cardamom ground	¼ tsp
8.	Ginger chopped	1 tsp
9.	Cooking oil	3 tbsp
10.	Tomato	2
11.	Coconut cream	1 cup
12	Coriander leaves	2 tsp

Method

1. Clean the prawns, leaving the tails intact.
2. Toss the prawns with the turmeric and keep aside.
3. Place the onion, garlic, paprika, red chillies, ginger and cardamom in a food processor and mix until a paste is formed.
4. Heat the oil in a deep-frying pan, carefully add the paste, stir it into the oil and cook over low heat for 10 minutes.
5. When the paste is cooked, it should be a golden brown colour.
6. Add prawns, tomatoes and coconut cream, stir well and simmer for about five minutes or until the prawns are cooked.

7. Stir in the coriander and season with salt.
8. Serve.

CU16) Goan Prawn Curry

This delicious Goan curry tastes great with plain steamed or boiled rice. It can also be made with fish or chicken if you so prefer.

Ingredients

1. Large prawns shelled and cleaned	1 kg
2. Red chillies	8–10
3. Pepper corns	6
4. Turmeric powder	½ tsp
5. Cumin seeds	1 tsp
6. Garlic	8 cloves
7. Ginger	1" piece
8. Tamarind	Golf ball size lumps
9. Onion thinly sliced	2 medium
10. Vegetable oil	3 tbsp
11. Salt	to taste

Method

1. Soak the tamarind in half a cup of hot water. Keep aside.
2. Heat a griddle or flat pan on a medium flame and gently roast (stirring often) the cumin seeds, peppercorns and dry red chillies till they begin to darken and release a faint aroma. Remove from fire and cool.

3. Heat the oil in a pan on a medium flame and add the onions. Fry till soft and slightly golden.
4. Remove from the fire and mix with the other ingredients (except prawns and tamarind) and grind to a smooth paste in a food processor.
5. Squeeze the soaked tamarind (in the water) with your hand to release the softened pulp. Do this till you have released all the pulp.
6. Strain and keep the juice aside.
7. Put the onion-spice paste back into the pan in which you fried the onions. Fry for 1 minute and add 1 ½ cups warm water and the tamarind juice. Bring to a boil on a medium flame. Simmer the flame.
8. Add the prawns and cook till done (they will turn from pink to opaque). Do not overcook as they will become rubbery.
9. Season with salt.
10. Serve piping hot with steamed or plain boiled rice or Sannas (steamed Goan rice cakes).

CU17) Kerala Prawn Curry

Ingredients

1.	Raw Prawns	500 g
2.	Vinegar	4 tbsp
3.	Dry Red Chillies	4 large
4.	Desiccated Coconut	6 tbsp
5.	Mustard Seeds	1 tsp
6.	Large Onion Sliced	1

7.	Vegetable Oil	4 tbsp
8.	Curry Leaves	10
9.	Fresh Ginger crushed	2 inch
10.	Garlic crushed	2 cloves
11.	Turmeric Powder	1 tsp
12.	Coriander Powder	1 tbsp
13.	Chilli Powder	1 tsp
14.	Tomatoes chopped	2
15.	Hot Water	125 ml
16.	Salt	to taste
17.	Coriander Leaves	to garnish

Method

1. Shell the prawns and de-vein them by lightly slitting back. Rub them with a little salt and 2 tsp of Vinegar and keep aside.

2. Dry roast the chillies, coconut, mustard seeds, onion and blend them to a fine paste to make a masala.
3. Heat Oil in a pan and fry the curry leaves. Reduce heat and add the ginger and garlic and fry for a few minutes until the raw smell appears.
4. Add the turmeric powder, ground coriander, chilli powder and the masala.
5. Stir-fry until oil separates. Add the tomatoes, salt to taste and the hot water.
6. Bring to boil and allow simmering for 5 minutes.
7. Drain the prawns, add them to the gravy and mix well. Add the remaining vinegar when the prawns begin to cook and turn pinky orange color.
8. Cook for a further two minutes. Do not over cook the prawns as they becomes hard.

CU18) Lobster Curry

Lobster meat cooked with yoghurt and spices to give this aromatic delicacy.

Cooking Time (approx.): 18 minutes

Ingredients

1.	Lobster shelled and meat washed	500 gm
2.	Big Onion	(Chopped) 1
3.	Tomato (chopped)	2 medium
4.	Coriander powder	2 tsp
5.	Cumin powder	2 tsp
6.	Turmeric powder	½ tsp

7.	Red Chilli powder	1 tsp
8.	Black pepper powder	½ tsp
9.	Garam Masala	½ tsp
10.	Yoghut	2 tbsp
11.	Oil	2 tbsp
12.	Water	4 tbsp
13.	Salt	to taste
14.	Grind to fine paste	
15.	Grated coconut	2 cups
16.	Garlic	2 flakes
17.	Ginger chopped	1 tsp
18.	Green chillies	2
19.	Poppy seeds	1 tsp

Finely chopped coriander to garnish.

Method

1. Heat the oil in a heavy-bottomed pan and saute the onions on medium heat for about 3 minutes or till the onions are light brown in color. Add the ground paste and fry till it is golden-brown. Add the tomato pieces and fry till the oil leaves the sides. Add all the spice powders except for the garam masala (hot spice mix) and mix well.
2. Mix in the lobster meat, salt and the yoghurt. Stir-fry on high heat briefly. Stir in the water and bring to a boil. Cover and simmer on low heat for about 15 minutes or till the lobster meat is cooked and tender. Sprinkle the garam masala (hot spice mix) and garnish with finely chopped fresh coriander.

Tip

☆ It is important that the lobster is fresh because its meat is easily removable then.

Serve hot with: steamed rice and/or slices of bread, rotis

Serves: 4

CU19) Lobster Pineapples Curry

Ingredients

1.	Rock lobster tails	1 (225 gm)
2.	Butter	2 tbsp
3.	Green ball pepper chopped	2 tbsp
4.	Chicken broth	¾ cup
5.	Curry powder	¼ tsp
6.	Can pineapple chunks drained	225 gm
7.	Chestnuts water	¼ cup
8.	Cashew nut	¼ cup
9.	Flour	2 tbsp

Method

1. Boil lobster tails according to package directions.
2. Cut away underside membrane and remove meat from shells.
3. Reserve shells for serving.
4. Cut meat into ½inch thick slices.
5. Melt butter in saucepan. Sauté green pepper in butter until tender, about 5 minutes.
6. Stir in flour. Gradually stir in broth and curry powder.

7. Cook over low heat, stirring constantly until mixture thickens.
8. Stir in pineapple chunks, water chestnuts, cashew nuts and lobster pieces.
9. Heat thoroughly.
10. Spoon mixture into shells.
11. Serve with seasoned rice.

CU20) Prawn Soup

Ingredients

1.	Prawns (medium size)	500 gm
2.	Turmeric	¼ tsp
3.	Ginger	¼ " piec
4.	Potato	1
5.	Onion	2
6.	Tomatoes Chopped	2
7.	Butter	1 tbsp
8.	Flour	1 tsp
9.	Pepper	to taste
10.	Salt	to taste.

Method

1. Wash prawns and boil with the turmeric, ginger, potato, one onion tomatoes and seasoning.
2. When the prawns are cooked, remove the juice.
3. Strain this juice along with the broth and set the prawns aside.
4. Melt the butter and fry the remaining chopped onion and the flour.

5. Cook for 2 minutes.
6. Pour in the prawns and boil for few minutes
7. Serve hot
8. Serves 6

CU21) Kerala Style Fish Curry

Ingredients

1.	Tuna fish cubes	600 gm
2.	Kokum	6 nos
3.	Red chilli powder	10 gm
4.	Coriander powder	10 gm
5.	Green chilli	4 nos
6.	Shallot paste	50 gm
7.	Dry coconut paste	100 gm
8.	Coconut oil	50 ml
9.	Tempering spices	1 recipe
10.	Salt	1 tsp

Method

1. Wash and soak with 400 ml of water
2. Heat oil and fry the tempering ingredients and remove.
3. In the remaining oil, fry coconut paste, Kokum pieces, green chilli and Shallot until they develop a roasted smell.
4. Add Kokum water, powdered ingredients, salt and fish.
5. Simmer until the fish has cooked completely.

6. Add the fried tempering ingredients.
7. Mix and serve.

Usually in Kerala State, India this curry is cooked in earthenware pots and served after a day.

CU22) Andhra Fish Curry (Seer Fish)

Fish simmered in a roasted onion and coconut paste with spices. The people of Andhra Pradesh love this spicy thick coconut gravy.

Cooking time (approx.): 16 minutes

Ingredients

1. Seer fish cut into pieces	500 gm
2. Garlic paste	1 tbsp
3. Tamarind paste	1 tbsp
4. Curry leaves	1
5. Lemon juice	1 tbsp
6. Ghee	2 tbsp
7. Salt	to taste
8. Coriander leaves finely chopped	To garnish

Grind to a Fine Paste

1. Coconut grated	1 cup
2. Red chillies dry	4
3. Onion	1
4. Coriander seeds	1 tbsp
5. Cumin seeds	1 tbsp
6. Turmeric powder	½ tsp

Method

1. Rub the washed fish pieces with some salt, lemon juice and a pinch of the turmeric powder. Keep aside for 5 minutes. Wash well. Smear some oil on the onion and roast it on a flame or oven till it is well-browned. Cool and chop.
2. In a pan, dry-roast the grated coconut on low heat till it is brown and aromatic. Cool and grind both to a fine paste along with the other ingredients given above.
3. Heat the ghee (clarified butter) in a heavy-bottomed pan and add the curry leaves. Add the ground paste and the garlic paste.
4. Fry on medium heat for about 4 minutes or till it is lightly browned and aromatic.
5. Add the fish pieces, tamarind paste and salt. Stir gently. Cover and cook on low heat for about 12 minutes or till the fish is cooked but firm. Garnish with finely chopped fresh coriander.

Tip

- ☆ Traditionally, ghee is a must for this rich and flavourful recipe. However, it may be replaced by any cooking oil.
- ☆ Serve hot with: steamed rice and/or warm Lacy Rice and Coconut Pancakes (Appam)

CU23) Hilsa Fish Curry with Mustard

It is a very very delicious unforgettable fish curry

Preparation time: 30 Minutes

Ingredients

1.	Hilsa Fish (Whole)	3 lb
2.	Mustard seeds	9 tbsp
3.	Green chilli	9
4.	Turmeric powder	1.5 tsp
5.	Red chilli powder	3 tbsp
6.	Salt	to taste
7.	Negella	0.5 tsp

Method

1. Take one whole hilsa fish and wash. After removing the head cut into steaks about ¾ inch thick. Remove the scales. Select 8 pieces. Rub 1 teaspoon of turmeric powder and 1 tea spoon of salt all over the pieces.

2. Grind to a fine paste the mustard seeds, 5 green chilies and ½ tea spoon of salt in an electric grinder. You will need to add water about ½ cup and it will take about 4 minutes, mustard seeds being together.
3. Mix red chili powder and turmeric powder with 3 table-spoon of water. For better taste, it is advisable to grind to a fine paste 10 or 12 whole red chilli with two table-spoon of water in a grinder
4. Heat oil in a non-stick thick bottom pan. Add kalonji and 4 green chillies and after about ½ minute add red chilli powder and turmeric powder mixture and fry with stirring for about 4 to 5 minutes. (If you do not want to be too hot then fry for a few minutes more adding little water)
5. Add the hilsa fish pieces one by one. Turn each piece. Add the mustard paste diluted with water (strain through a strainer so that only the juice is used) and two green chilli sliced and bring to boil adding ½ cup of water. Reduce heat, simmer for about ten minutes till gravy thickens and continue frying carefully turning the pieced till gravy becomes little dry.
6. Serve hot with basmati rice.

CU24) Brinjal Prawn Curry

Brinjal Prawn Curry is a delightful Malabar recipe. Learn how to make/prepare Brinjal Prawn Curry by following this easy recipe.

Ingredients

1.	Prawns	2 cups

2.	Brinjal	1 cup
3.	Tomato (Chopped)	½ cup
4.	Cumin seeds	1 tbsp
5.	Mustard seeds	1 tbsp.
6.	Turmeric powder	½ tbsp
7.	Green chillies	4
8.	Garlic flakes	4
9.	Water	½ cup
10.	Oil	for frying
11.	Salt	to taste

Method

1. Cut brinjal into slices. Grind mustard seeds, green chillies, coriander leaves and garlic flakes to make a paste by adding water.
2. Heat the oil. Fry brinjal slices till crisp. Keep it aside.
3. Now fry prawns till golden brown and keep them aside.
4. Add cumin and mustard seeds heated oil. Now add chopped tomatoes.
5. Cook them for 2 minutes.
6. Mix fried prawns and brinjals.
7. Stir them well. Add prepared paste, turmeric powder and salt.
8. Cook the mixture for 5 minutes.
9. Add ½ cup water. Cook for 5 minutes. Brinjal Prawn curry is ready.

Chapter 7

Canned Seafood Products

Ca1) Salmon Canned Fish Cachets

Ingredients

A) Filling

1. One tin of any fish (Salmon preferred)	
2. Olive oil	2 tbsp
3. Grated cheese	1 desert spoon
4. Green chillies	2
5. Chilli powder	½ tsp
6. Mustard	½ tsp
7. Pepper	1.4 tsp
8. Salt	to taste

B) Pastry

1. Flour	2 cups
2. Soojee	1 cup

3. Butter	2 cup
4. Salt	to taste
5. Pepper	to taste
6. Eggs	3
7. Milk	little

C) Sauce

1. Yolk of hard cooked egg	1
2. Vinegar	little
3. Oil	1 desert spoon
4. Sugar	1 tsp

Method

A) For Filling

1. Put the oil in a pan and when hot, add the fish, cheese, green chillies, chilli powder.
2. Seasoning the gravy.
3. Cook till gravy dries.
4. Set aside.

B) For Pastry

1. Mix the flour, soojee and salt with butter till the texture resembles bread crumbs
2. Add the egg (Yolk only) and milk to form soft dough.
3. Kneel well and either from into individual cachets
4. Line a large dish with the pastry and back in a hot oven.

C) For Sauce

1. Mix the yolk of the hard cooked egg with the vinegar to form a paste.

2. Add seasoning, oil and sugar.
3. Mix till the sugar dissolves.

Place the fish on to the backed shell. Garnish with tomatoes and whites of the hard cooked egg, and spoon the sauce over.

Serves 6

Ca2) Terrific Tuna Terazzini

This is a quick-to-make weekday standby. Chances are you'll have most of the ingredients on hand and you can pull it together in a flash. There's no time-consuming butter-and-flour thickener and light sour cream keeps the calories down.

Ingredients

1. Flaked Light Tuna, drained	1 can (170 gm)
2. Spaghetti	125 gm
3. Frozen peas	1 cup
4. Light sour cream	1 cup
5. Mustard	½ tsp
6. Sherry	2 tbsp
7. Grated cheese	1 cup
8. White pepper	pinch

Method

1. Preheat oven to 375°F (190°C).
2. Bring a large pot of salted water to a full rolling boil.
3. Add the spaghetti and boil uncovered, stirring occasionally, until cooked al dente, about 8 to 10 minutes.
4. Then, drain but do not rinse.
5. Meanwhile, rinse the peas under running water to remove the ice crystals.
6. Drain well and set aside.
7. In a large bowl, stir the sour cream with the mustard and sherry, if using.
8. Add the drained cooked spaghetti and stir until coated.
9. Then stir in ½ cup mozzarella cheese.
10. Add the tuna and peas. Stir just until evenly distributed.
11. Taste and add more mustard or some white pepper, if you wish.
12. Turn the tuna mixture into a deep-dish pie plate. Sprinkle with the remaining cheese.
13. Bake about 25 minutes in the center of the preheated oven until hot.
14. Serve 6.

Recipe courtesy of "The New Chatelaine Cookbook" by Monda Rosenberg, Maclean Hunter, p.174

Ca3) Chinese Sweet and Sour Salmon

Ingredients

1.	Skinless boneless Salmon Can	3 X 170 gm can
2.	Medium egg noodles	250 gm
3.	Orange juice	1 orange
4.	Clear honey	4 tbsp
5.	Rice vinegar	2 tbsp
6.	Soy sauce	4 tbsp
7.	Tomato ketchup	4 tbsp
8.	Vegetable oil	2 tbsp
9.	Fresh fried vegetables	400 gm
10.	Unsalted cashew nuts	50 gm
11.	Sesame seeds	2 tsp

Method

1. Drain the salmon and break into large chunks.
2. Set aside

3. Put the egg noodles into a heatproof bowl and pour over enough boiling water to cover.
4. Leave to soak for 6 minutes.
5. Mix together the orange juice, honey, vinegar, soy sauce, tomato ketchup and corn flour in a small jug.
4. Heat the vegetable oil in a wok or large frying pan.
5. Add the vegetables and stir fry for 2 - 3 minutes.
6. Drain the noodles thoroughly and add them to the vegetables.
7. Stir the orange juice mixture and add to the wok.
8. Stir and cook for another 2 - 3 minutes until piping hot and slightly thickened.
9. Just before serving, stir the salmon chunks and cashew nuts through the mixture.
10. Serve, sprinkle with sesame seeds.

Ca4) Crabmeat on Toast

A simple variation of this recipe will result in yet another tasty treat. Substitute the can of crabmeat for a can of shrimp meat and replace the tomatoes with thin slices of avocado. This version does not need to be cooked.

Ingredients

1. Crabmeat Can well drained	1 can (120 gm)
2. Mayonnaise	¼ cup
3. Lemon juice	½ tsp
4. Stalk celery, chopped	½
5. Green onion, chopped	1 tbsp

6.	Tomato (very thin sliced)	1
7.	Toasted bread	2 slices
8.	Fresh ground black pepper	pinch

Method

1. In a bowl mix together the crabmeat, mayonnaise, lemon juice, celery and green onions.
2. Place 4 slices of tomato on each piece of toasted sourdough.
3. Top with the crabmeat mixture and fresh ground black pepper to taste.
4. Place under the broiler for 2 to 3 minutes until bubbly and warmed through.

Recipe courtesy of the "Tin Fish Gourmet" cookbook by Barbara-jo McIntosh.

Ca5) Canned Tuna and Crab Salad

Ingredients

1.	White Tuna can	375 gm can
2.	Eggs	2
3.	Crabmeat can, shredded	186 gm can
4.	Sour cream	1 tsp
5.	Yellow mustard	1 tsp
6.	Dried dill weeds	½ tsp
7.	Lemon pepper	1/8 tsp
8.	Dill pickle relish	2 tbsp
9.	Sweet pickle relish	2 tbsp
10.	Onion finely chopped	¾ cup

Method

1. Place eggs in a saucepan and cover with cold water.
2. Bring water to a boil and immediately remove from heat.
3. Cover and let eggs stand in hot water for 10 to 12 minutes.
4. Remove from hot water, cool, peel and chop finely.
5. In a mixing bowl, combine tuna, crabmeat, mayonnaise, sour cream and mustard.
6. Stir in chopped egg, dill weed, lemon pepper, dill and sweet relish and chopped onion.
7. Mix thoroughly.

Chapter 8

Minced Seafood Products

M1) Fish Balls from Kheema

Ingredients

1.	Minced meat	1 kg
2.	Chilli powder	30 gm
3.	Coriander powder	10
4.	Turmeric powder	1 tsp
5.	Double refined oil	150 gm
6.	Big Onion	180 gm
7.	Egg	1
8.	Bread powder	30 gm
9.	Aginomotto	½ tsp
10.	Salt	To taste

Method

1. Thaw the minced meat and keep aside for the water to drain.
2. Cut the onion into very small pieces, fry it in double refined oil, and mix chilli powder, coriander powder, turmeric powder etc.
3. Mix the minced meat with the fried onion and masala mixture, followed by the addition of beaten eggs, bread powder, salt, aginomotto etc. and mix homogeneously.
4. Form the about preparation into balls of about 3 cm diameter and steam for 10–15 minutes.
5. Serve hot.

M2) Fish Kabab

Ingredients

1. Keema (Minced meat)	1 kg
2. Channa Dal	250 gm
3. Garlic	15–20 pods
4. Cardamom	6
5. Jeera	1 tsp
6. Cinnamon	Two 1″ pieces
7. Ginger	2" piece
8. Onion	2 large ones
9. Pepper corns	8 nos
10. Red Chillies	10 nos
11. Cloves	8
12. Salt	To taste

13.	Eggs	2
14.	Green chillies	6 nos.

Method

1. Grind ingredients 1–11. Add 2 glasses of water and boil.
2. Simmer till water is completely absorbed and kheema is tender.

To make Kabab

1. Beat 2 eggs and mix chopped onion and green chillies.
2. Shape into small flattened balls or Kababs.
3. Fill with finely chopped onion and green chillies
4. Deep fry till golden brown.
5. Serve hot with chutney and lemon twists.

M3) Fried Fish Fingers

Ingredients

1.	Frozen minced meat	1 block of 450 gm
2.	Flour	40 gm
3.	Red Chilli powder	½ tsp
4.	Egg	2 no
5.	Bread crumbs	40 gm
6.	Salt	To taste.
7.	Double refined oil	for frying.

Method

1. Cut the frozen minced meat into 2x8 cm pieces.
2. Coat thinly with a mixture of flour, salt and chilli powder.

3. Dip in lightly beaten egg.
4. Coat with bread crumbs.
5. Shake of excess crumbs and deep fry in hot oil .
6. Drain oil on a brown or tissue paper.
7. Serve hot with fried potato chips and garnish with tomato and lemon twists.

M4) Fish Cutlets

Ingredients

1.	Minced meat–Kheema	1 kg
2.	Potato	500 gm
3.	Onion	300 gm
4.	Ginger	50 gm
5.	Green chilli.	40 gm
6.	Curry leaves	few
7.	Turmeric powder	1 tsp
8.	Pepper powder	1 tsp
9.	Chilli powder	100 gm
10.	Spices mixture	40 gm
11.	Ground nut oil	80 gm
12.	Egg	1
13.	Bread powder	100 gm
14.	Maida	50 gm
15.	Salt	20 gm

Method

1. Cook the minced meat with salt and turmeric powder.

2. Cook the potato, peel, smash it and mix with the cooked meat.
3. Slice the onion and green spices (4 to 6),
4. Mix it with pepper powder, chilli powder and spices mixture.
5. Mould it according to required sizes.
6. Batter it in the battering mixture containing egg, water and Maida.
7. Bread the battered cutlets and deep fry in oil
8. Serve hot.

M5) Fish Cakes

Ingredients

1.	Minced meat	1 kg
2.	Starch	125 gm
3.	Vegetable oil	70 gm.
4.	Pepper powder	1 tsp
5.	Galic peeled	5–6 cloves
6.	Green chillies	10 gm
7.	Mint leaves	5 gm
8.	Butter	100 gm
9.	Breading Mixture	120 gm
10.	Salt	To taste

Method

1. Grind the minced meat with salt, starch powder and spices.
2. Mould it in metallic or wooden moulds of different shapes.

3. The thickness must not exceed 1 cm.
4. Cool the moulded fish cakes in a freezer till it becomes semi frozen.
5. Make the batter by mixing wheat powder or starch with chilled water in the ratio 1 : 1.
6. Breading mixture is made by powdering dried bread.
7. Dip the semi frozen fish cake in the batter and spread the bread powder uniformly over it.
8. Freeze the battered and breaded fish cakes.
9. Fry the frozen fish cake in hot oil till it attains a golden yellow colour.

Appendices

Table 1: List of Various Herbs and Species with their Hindi Names

Sl.No.	*English Names*	*Hindi Names*
1.	Almond	*Badam*
2.	Asafoetida	*Hing*
3.	Bay Leaves	*Tej patta*
4.	Bay salt	*Marine (Raw) salt.*
5.	Beetroot	*Chukandar*
6.	Bengal Gram Oal	*Chana Oal*
7.	Broth	*Fish soup*
8.	Butter milk	*Chhas*
9.	Capsicum	*Shimala Mirch*
10.	Cardamom pods	*Elaichi dana*
11.	Cardamom	*Elaichi*
12.	Carom seeds	*Ajwain*
13.	Carrot	*Gajar*
14.	Cashewnut	*Kaju*
15.	Cashiya leaf	*Tamalpatra*
16.	Chilli	*Mirchi*
17.	Cinnamon	*Oalchini*
18.	Cloves	*Laung*
19.	Coconut	*Khopra, Narial*
20.	Condiment us	*Masala*
21.	Coriander seed	*Saboot dhania*
22.	Coriander seeds, ground	*Ohania powder*
23.	Coriander, Fresh	*Hara Ohania*
24.	Cumin seeds, Black	*Shah Jeera*
25.	Cumin seeds, white	*Jeera*

Contd...

Table 1–Contd...

Sl.No.	*English Names*	*Hindi Names*
26.	Curd	*Oahi*
27.	Curry Leaves	*Kari patta*
28.	Egg	*Anda*
29.	Fennel	*Saunf*
30.	Fenugreek seeds	*Methi dana*
31.	Fenugreek seeds, dried	*Kasoori methio*
32.	Fresh green Chillies	*Hari Mirch*
33.	Garam Masala: A Spice Blend	*Garam Masala*
34.	Garlic	*Lasun*
35.	Ginger	*Adrak*
36.	Goraka	*Anardana*
37.	Green pepper	*Lila Mari*
38.	Lime	*Nimbu*
39.	Mace	*Javitri*
40.	Mango Powder	*Amchoor*
41.	Melon seeds	*Magaz*
42.	Milk	*Dudh*
43.	Mint	*Poodina*
44.	Mustard seeds	*Rai, Sarson*
45.	Nigella Seeds	*Kalaunp*
46.	Nutmeg	*Jaiphal*
47.	Onion	*Pyaz*
48.	Pepper	*Kali Mirch*
49.	Peppercorns	*Saboot Kali Mirch*
50.	Pomegranate seeds, Dried	*Anardana*
51.	Pomfret	*Paplet*
52.	Potato	*Alu*
53.	Prawn	*Jhinga*
54.	Raisin	*Kismis*

Contd...

Table 1–Contd...

Sl.No.	*English Names*	*Hindi Names*
55.	Red Chilli powder	*Lal Mirch*
56.	Red Peppercorns, Dried	*Sukhi lal Mirch*
57.	Refined Flour	*Maida*
58.	Saffron	*Kesar*
59.	Semolina	*Soji*
60.	Sesame Seeds	*Til*
61.	Spring Onion	*Uli Dungri*
62.	Sugar	*Chini*
63.	Tamarind	*Emali*
64.	Tomatoes	*Tamater*
65.	Turmeric	*Haldi*
66.	Vinegar	Sirka
67.	Yogurt	Dahi

Table 2: Standard Weight and Measures

Abbreviation Used	*Meaning*	*Measures*
tsp	Teaspoon	5ml
tbsp	Table spoon	3 tsp = 15 ml
dsp	Desert spoon	
Gm/gms	Gram	15.432 grain
Cup (Dried product)	Cup	16 tbsp
Cup of milk	Cup	250ml
Kg	Kilogram	1000 gm
1 Ounce		28.35 gm

Table 3: Glossary of Selected Cookery Terms

Batter.	Any mixture of dry ingredients and liquid that is stirred or beaten and can be poured.
Blending.	Combining ingredients with a spoon, beater etc. to achieve a uniform mixture
Boiling.	Cooking in liquid at a temperature of 100° C (212° F)
Chilling.	Cooling food, without freezing it, in the refrigerator.
Curd.	Semi solid part of milk produced by souring.
Deep-frying.	Frying food by immersing it in hot fat or oil.
Dressing.	1. Sauce for a salad
	2. Stuffing for meat or poultry.
Freezing.	Solidifying or preserving food by chilling and storing it at below 0° C
Garnishing.	Decorating with portions of colourful and contrasting food.
Grating.	Shredding foods by rubbing against a grater
Gravy.	1. Juice exuded by roasted meat and poultry.
	2. Sauce made from these juices by boiling with stock or wine and sometimes thickened with flour.
	3. Thick liquid made by adding water to vegetables and/ or masala.
Saute:	To fry food rapidly in shallow, hot fat, tossing and turning it until evenly browned.
Simmering.	Cooking in liquid which is heated to just below boiling point.
Stirring.	Mixing with a circular movement, using a spoon or fork.
Vinegar.	A clear liquid, consisting of chiefly acetic acid, obtained by the fermentation of wine, cider or malt beer.

www.ingramcontent.com/pod-product-compliance
Ingram Content Group UK Ltd.
Pitfield, Milton Keynes, MK11 3LW, UK
UKHW020207290926
15470UKWH00001B/43

9 788170 359937